Exploring Evolution from Scientific and Biblical Perspectives

◀ *I am quite conscious that my speculations run beyond the bounds of true science... It is a mere rag of a hypothesis with as many flaw(s) and holes as sound parts.* ▶

Charles Darwin in 1887

Life and Letters, Volume 2

Exploring Evolution from Scientific and Biblical Perspectives

Andy Williams

Published by Graceful Script

Available for purchase on Amazon.com

First Printing: February 2016

Careful attention has been made to correctly state quotations and provide the source of each quotation.

Published by Graceful Script

ISBN: 978-0692624005
Cover design by GaBrielle Boyd

Published and printed in the United States of America

⌘ Contents ⌘

Dedication 8
Preface 9
The Layout and Features of this Book...10
Introduction 11

Chapter 1 – Solar System and Universe 15
Scientific Reason 1: Big Bang 16
Scientific Reason 2: Laws of Planetary Motion 19
Scientific Reason 3: Spiral Galaxies 20
Scientific Reason 4: The Poynting-Robertson Effect 21
Scientific Reason 5: Our Shrinking Sun 22
Scientific Reason 6: Heat Radiation of Jovian Planets 24
Scientific Reason 7: Short-Lived Comets 24
Scientific Reason 8: Supernovas 25

Chapter 2 – Physical Science 30
Scientific Reason 9: Matter 31
Scientific Reason 10: Atoms and Elements 32
Scientific Reason 11: Water 34
Scientific Reason 12: Universal Constants 35
Scientific Reason 13: Laws of Physics 37
Scientific Reason 14: Second Law of Thermodynamics 38
Scientific Reason 15: Formation of Stalactites, Diamonds, and Coal 39
Scientific Reason 16: Formation of Granite 40

Chapter 3 – Earth 45
Scientific Reason 17: Anthropic Principle 46
Scientific Reason 18: Oxidizing Atmosphere 47
Scientific Reason 19: Atmospheric Helium 48
Scientific Reason 20: Earth's Magnetic Field 49
Scientific Reason 21: Earth's Rotation and Orbit 50
Scientific Reason 22: Cycles 51
Scientific Reason 23: Rapid and Catastrophic Evidence 52
Scientific Reason 24: Mississippi River Delta 55
Scientific Reason 25: Ocean Sodium 55
Scientific Reason 26: Ocean Sediment and Mud 56

Chapter 4 – Fossils 61
Scientific Reason 27: Fossils in Sedimentary Layers 62
Scientific Reason 28: Number of Transitional Human Fossils 65
Scientific Reason 29: Human Artifacts in the Fossil Record 67

Scientific Reason 30: Cambrian Explosion 67
Scientific Reason 31: Direct Dinosaur Ancestor Fossil 69
Scientific Reason 32: "Living" Fossils 70
Scientific Reason 33: Polystrate Fossils 71
Scientific Reason 34: Petrification 72
Scientific Reason 35: Soft Tissues in Fossils 72

Chapter 5 – Geology 78
Scientific Reason 36: Geologic Column 79
Scientific Reason 37: Dating of Geologic Layers 80
Scientific Reason 38: Flat and Horizontal Sedimentary Layers 81
Scientific Reason 39: Mount Saint Helens 82
Scientific Reason 40: Oil Creation and Pressure 83
Scientific Reason 41: The Grand Canyon 85

Chapter 6 – Radiometric Dating 90
Scientific Reason 42: Radiometric Dating Credibility 91
Scientific Reason 43: Carbon-14 Radiometric Dating 93
Scientific Reason 44: Assumptions in Carbon-14 Dating 94
Scientific Reason 45: Carbon-14 SPR Versus SDR 96

Chapter 7 – Living Things 99
Scientific Reason 46: Spontaneous Generation 100
Scientific Reason 47: Law of Biogenesis 101
Scientific Reason 48: Creating Proteins 102
Scientific Reason 49: Odds of the Formation of Enzymes 103
Scientific Reason 50: Cells 105
Scientific Reason 51: Bacterium Flagellum 107
Scientific Reason 52: Butterfly Metamorphosis 108
Scientific Reason 53: Animal Design, Beauty, and Amazing Capabilities 109
Scientific Reason 54: Patterns of Discontinuity 111
Scientific Reason 55: Irreducible Complexity 112
Scientific Reason 56: Photosynthesis 114
Scientific Reason 57: Seeds 115
Scientific Reason 58: Fermi Paradox 116

Chapter 8 – Information and DNA 122
Scientific Reason 59: Information 123
Scientific Reason 60: Laws of Math 125
Scientific Reason 61: Fibonacci Sequence 126
Scientific Reason 62: Design, Beauty, and Complexity of DNA 128
Scientific Reason 63: DNA Mutations 131

Chapter 9 – People 137
Scientific Reason 64: Growth Rate of the Human Population 138
Scientific Reason 65: Amazing Design, Beauty, and Capabilities in People 139
Scientific Reason 66: The Energy Currency of Life 142
Scientific Reason 67: Vestigial Organs 143

Chapter 10 – Evolution 147
Scientific Reason 68: Existence of Non-Material Things 148
Scientific Reason 69: Naturalistic Assumption 149
Scientific Reason 70: Statistical Impossibility 150
Scientific Reason 71: Natural Selection 151
Scientific Reason 72: Speciation 153
Scientific Reason 73: Haeckel's Biogenetic Law 153
Scientific Reason 74: Extinction of Dinosaurs 155
Scientific Reason 75: Homologous Structures 156
Scientific Reason 76: Morphological Tree Diagrams 157
Scientific Reason 77: No Conclusive Evidence 159

Chapter 11 – Biblical Reasons 167
Biblical Reason 1: Order of Creation Events 168
Biblical Reason 2: Comparing Worldviews 169
Biblical Reason 3: Foundation Attack 172
Biblical Reason 4: Naturalism's Failures 173
Biblical Reason 5: Theistic Evolution 174
Biblical Reason 6: Gap Theory 175
Biblical Reason 7: Writing Style of Genesis 176
Biblical Reason 8: Clear Meaning of Genesis Chapters 1 and 2 177
Biblical Reason 9: View of People 179
Biblical Reason 10: Genesis Genealogies 180
Biblical Reason 11: Created Kinds 183
Biblical Reason 12: Thorns 183
Biblical Reason 13: View of Death 184
Biblical Reason 14: Future Perfection 186
Biblical Reason 15: Perspective of All New Testament Writers 187
Biblical Reason 16: Old Testament Verses 188
Biblical Reason 17: New Testament Verses 191
Biblical Reason 18: Personal Perspective of the Lord Jesus Christ 193
Biblical Reason 19: Authority of the Lord Jesus Christ's Word 196

Chapter 12 – Why Does It Matter 204

Appendix A: Summary List of 77 Scientific and 19 Biblical Reasons 209
Appendix B: Resources For Your Own Exploring 221
Index 223

⌘ Dedication ⌘

This book is dedicated to two groups of people:

I dedicate this book to the millions of people who have been misled into thinking that evolution over billions of years qualifies as true science.

I dedicate this book to the millions of young people who have lost faith in the Lord Jesus Christ because you were never taught the scientific and Biblical answers to the beliefs promoted by evolution and billions of years.

If my years researching and writing this book give you a new understanding into the beliefs of evolution and billions of years, it will have been time well spent.

⌘ Preface ⌘

I have enjoyed studying science since my college years. Attending Concordia College in Portland, Oregon, I double majored in Life and Natural Sciences and triple minored in Biblical Studies, Education, and Computer Applications. My science classes sparked an interest in the ongoing debate about human origins. That debate boils down to two options: 1) We are the result of evolution over billions of years as suggested by Charles Darwin, or 2) We are the result of special creation about 6,000 years ago as described in the Bible.

Following my college years in the 1980s, I began building a list of reasons why evolution is scientifically and biblically impossible. I have always enjoyed solving problems and answering questions such as "Why does that happen?" or "How does that work?" Building upon my college education, my 20-year career at Microsoft shaped and polished my deductive thinking and problem-solving skills. So in response to constantly reading and hearing the assumption that evolution over billions of years is "scientific" and the Biblical creation event over seven days is "just religion", this book emphasizes 77 scientific reasons complemented by 19 Biblical reasons.

Let me be clear about what evolution is and what it is not... Some people define evolution merely as "change" and you may hear someone say, "That paper evolved from 1 page to a 10 page document". Others may think that evolution occurs merely when a species changes, such as when an animal grows different fur length in response to seasonal changes in its environment. Neither of these fits the thorough definition of evolution.

Rather, evolution is defined as the idea that random-chance, purposeless, materialistic, naturalistic, unintelligent, goalless, death-based, designless processes caused life to form billions of years ago, and these processes, following a succession of continual genetic and physical advancements over billions of years, produced all life on earth, including people.

It will be key to remember the definition of evolution as you read and consider the reasons in this book. As you read these 77 scientific and 19 Biblical reasons, I invite you to ask yourself: "Is evolution over billions of years actually scientifically or Biblically possible?"

⌘ The Layout and Features of this Book ⌘

- The first 10 chapters detail 77 scientific reasons why evolution over billions of years is impossible.
- Chapter 11 details 19 Biblical reasons why evolution over billions of years is impossible.
- At the end of chapters 1 to 11, endnotes and resources for the reasons included within that chapter are listed.
- The final chapter, chapter 12, highlights why it matters what you believe about evolution and billions of years.
- Where quotations are included, they are bracketed with speaker symbols.
- As a convenience, Appendix A lists all 77 scientific and 19 Biblical reasons together.
- Appendix B lists a collection of book and online resources researched and studied to write this book. These are valuable resources for your own exploring!
- An index offers key topics and noted people.

⌘ Introduction ⌘

Although many people consider it a scientific theory, according to the established scientific guidelines for determining a theory, evolution does not lend itself to the "criterion of falsifiability" and therefore does not actually qualify as a scientific theory. Said more simply: it cannot be tested. In the scientific community, in order for a theory to be scientifically viable, it must be testable. Scientific research involves making an observation, then making a hypothesis, then testing to see if the hypothesis is correct or not, and then offering a theory based on the results of the testing. This process has been coined the "scientific method".

The scientific method requires science to be observable, demonstrable, testable, repeatable, and after all of these steps, have the possibility of being refutable. Evolution fails to fulfill any of the scientific method's steps. Charles Darwin, the man we recognize for proposing evolution, made this eye-opening statement in 1887 about the evolutionary ideas he had proposed in his famous 1859 book *On the Origin of Species by Means of Natural Selection, or the Preservation of Favoured Races in the Struggle for Life*:

◀ *I am quite conscious that my speculations run beyond the bounds of true science... It is a mere rag of a hypothesis with as many flaw(s) and holes as sound parts.* ▶ [1]

Notice that Charles Darwin himself calls his ideas mere speculations that run beyond the known guidelines of real science. By acknowledging that his idea of evolution is a "rag of a hypothesis with as many flaw(s) and holes as sound parts", he too recognizes that evolution fails to qualify as a legitimate scientific theory. Tragically, Charles Darwin also notes the destructive impact to a society which believes and acts upon his ideas:

◀ *To have the unthinking masses accept all that I say would be calamity...* ▶ [2]

It is significant to note that evolution is based upon many unknowns that cannot be scientifically observed or substantiated:

- Unknown original chemicals which we cannot confirm
- Unknown life-forming processes that no longer exist
- Unknown life forms that cannot be found or confirmed
- Unknown reproductive methods that we know nothing about
- Unknown oceanic soup for which we have no evidence
- Unknown time in history that cannot be validated

Paleontologist and evolutionist George Gaylord Simpson, Ph.D., reminds us that observation is a requirement for ideas to be truly scientific:

◀ *It is inherent in any definition of science that statements that cannot be checked by observation are not really about anything... or at the very least, they are not science.* ▶ [3]

Zoologist and evolutionist Michael Rose, Ph.D., highlights that evolution is not scientific but rather represents a form of religion that opposes Christianity:

◀ *Evolution is promoted by its practitioners as more than mere science. Evolution is promulgated as an ideology, a secular religion - a full-fledged alternative to Christianity, with meaning and morality. I am an ardent evolutionist and an ex-Christian, but I must admit that in this one complaint - and Mr. [Duane] Gish is but one of many to make it - the literalists are absolutely right. Evolution is a religion. This was true of evolution in the beginning, and it is true of evolution still today... Evolution therefore came into being as a kind of secular ideology, an explicit substitute for Christianity.* ▶ [4]

Physicist and creationist Philip Johnson, Ph.D., summarizes why evolution is not actually scientific:

◀ *1) Evolution is grounded not on scientific fact, but on a philosophical belief called naturalism; 2) The belief that a large body of empirical evidence supports evolution is an illusion; 3) Evolution is itself a religion; and 4) If evolution were a scientific hypothesis based on rigorous study of the evidence, it would have been abandoned long ago.* ▶ [5]

Biologist and evolutionist Richard Lewontin, Ph.D., identifies a bias which some scientists embrace regarding the belief in evolution and its billions-of-years premise, in spite of its failure to be an actual scientific theory:

◀ *Our willingness to accept scientific claims that are against common sense is the key to an understanding of the real struggle between science and the supernatural. We take the side of science in spite of the patent absurdity of some of its constructs, in spite of its failure to fulfill many of its extravagant promises of health and life, in spite of the tolerance of the scientific community for unsubstantiated just-so stories, because we have a prior commitment, a commitment to materialism. It is not that the methods and institutions of science somehow compel us to accept a material explanation of the phenomenal world, but, on the contrary, that we are forced by our apriori adherence to material causes to create an apparatus of investigation and a set of concepts that produce material explanations, no matter how counter-intuitive, no matter how mystifying to the uninitiated. Moreover, that materialism is an absolute, for we cannot allow a Divine Foot in the door.* ▶ [6]

Biologist and evolutionist professor Paul Ehrlich, Ph.D., identifies why belief in evolution and its billions-of-years premise falls outside the realm of real science:

◀ *Our theory of evolution has become, as Popper described, one which cannot be refuted by any possible observations. Every conceivable observation can be fitted into it. It is thus 'outside of empirical science' but not necessarily false. No one can think of ways in which to test it. Ideas, either without basis or based on a few laboratory experiments carried out in extremely simplified systems, have attained currency far beyond their validity. They have become part of an evolutionary dogma accepted by most of us as part of our training.* ▶ [7]

Mathematician and evolutionist Wolfgang Smith, Ph.D., reminds us that not one shred of conclusive scientific evidence exists for evolution:

◀ *We are told dogmatically that evolution is an established fact; but we are never told who has established it, and by what means. We are told, often enough, that the doctrine is founded upon evidence, and that indeed this evidence 'is henceforward above all verification, as well as being immune from any subsequent contradiction by experience'; but we are left entirely in the dark on the crucial question wherein, precisely, this evidence consists. And the salient fact is this: if by evolution we mean macroevolution (as we henceforth shall), then it can be said with the utmost rigor that the doctrine is totally bereft of scientific sanction. Now, to be sure, given the multitude of extravagant claims about evolution promulgated by evolutionists with an air of scientific infallibility, this may indeed sound strange. And yet the fact remains that there exists to this day not a shred of bona fide scientific evidence in support of the thesis that macroevolutionary transformations have ever occurred.* ▶ [8]

Biologist and evolutionist George Wald, Ph.D., shares that evolution is based on the idea that the impossible can become possible with enough time:

◀ *The important point is that since the origin of life belongs in the category of at-least-once phenomena, time is on its side. However improbable we regard this event, given enough time it will almost certainly happen at least once... Time is in fact the hero of the plot. The time with which we have to deal is of the order of two billion years. What we regard as impossible on the basis of human experience is meaningless here. Given so much time, the 'impossible' becomes possible, the possible probable, and the probable virtually certain. One has only to wait: time itself performs miracles.* ▶ [9]

According to the established principles that define science and its scientific method, as determined by the scientific community for centuries, evolution does not qualify as real science, it does not fulfill the requirements of the scientific method, and it does not qualify as a scientific theory. Many evolutionists, creationists, and even Charles Darwin himself openly recognize and understand this.

Introduction Endnotes

1 – Charles Darwin, *Life and Letters,* Vol. 2 (1887) page 229.
2 – Charles Darwin, *Little Journeys to the Homes of the Great Scientists*, Editor Elbert Hubbard, written in a letter to Dr. Asa Gray, 1905.
3 – Frank Sherwin, "Science vs. Macroevolution", *Acts & Facts*, Vol. 1 (2001) page 30.
4 – Michael Ruse, "How evolution became a religion: creationists correct?" *National Post* (13May2000) pages B1, B3, and B7.
5 – Philip Johnson, *In Six Days – Why 50 Scientists Choose to Believe in Creation,* Editor John F. Ashton (Green Forest, AR: Master Books, 2001) page 308.
6 – Richard Lewontin, "Billions and Billions of Demons", *The New York Review of Books* (9January1997) Web 15September2015. <http://www.nybooks.com/articles/archives/1997/jan/09/billions-and-billions-of-demons>.
7 – Paul Ehrlich and L. Charles Birch, "Evolutionary history and population biology", *Nature*, Vol. 214 (22April1967) page 352.
8 – Wolfgang Smith, *Teilhardism and the New Religion* (Rockford, Illinois: Tan Books & Publishers, Inc., 1988) pages 2 and 5.
9 – George Wald, *The Origin of Life: The Physics and Chemistry of Life* (New York: Simon & Schuster, 1955) page 12.

⌘ Chapter 1 ⌘

Solar System And Universe

The universe we see when we look out to its farthest horizons contains a hundred billion galaxies. Each of these galaxies contains another hundred billion stars. That's $10^{(22)}$ *stars all told. The silent embarrassment of modern astrophysics is that we do not know how even a single one of these stars managed to form.* [3]

⌘ Scientific Reason 1 ⌘ Big Bang

Everyone has heard of the Big Bang. It is the top vote-getter in the scientific community in terms of proposing where everything in our universe originated. Briefly defined, the Big Bang proposes that billions of years ago our universe began in space as an incredibly small, extremely hot, highly energetic point called a singularity, and this singularity exploded and led to the creation of everything that exists today.

According to the principles of observational science, the Big Bang fails as a scientific theory for the following 10 reasons:

- The Big Bang cannot explain where matter and energy came from in the first place, in order to allegedly explode.

- We live in an existence based upon space and time. The Big Bang cannot explain how space and time originated.

- The Big Bang proposes that all matter and energy originally existed within an object called a singularity that was smaller than an atom. It certainly is not scientific to suggest that our universe, which has an estimated 200 billion galaxies each hosting an estimated 100 billion stars, could have originated in an object smaller than the tip of a pen.

- The Big Bang suggests that nothingness was densely packed together in a vacuum and this caused the singularity to explode. But there is no way to scientifically explain how nothingness became packed, what force packed it together, and how a vacuum (which is completely nothing) could have any amount of mass or density to even be packed.

- The Big Bang proposal is built on the belief in an original explosion, but explosions require heat, energy, and something to ignite the explosion. There is no scientific way to explain where the initial heat, energy, and ignition originated.

- The First Law of Thermodynamics tells us the energy in our universe cannot be created or destroyed and that the total amount of energy in our universe will always stay the same. The Second Law of Thermodynamics tells us that this constant amount of energy is moving from a more orderly and more usable state into a less orderly and less usable state. Scientists have coined the second law the "principle of entropy". Although the total amount of energy in our universe never changes (the First Law), the amount of usable energy decreases (the Second Law). Neither of these laws has ever been shown to be false, and all laws of physics are completely consistent with both laws. The Big Bang contradicts the First Law of Thermodynamics by suggesting that mass and energy were somehow mysteriously created in an initial explosion, and it contradicts the Second Law of Thermodynamics by suggesting that, along with evolution, everything has always been progressing from a less orderly and usable state to a more complex and orderly state.

- The Big Bang proposes that a mysterious "dark matter" surrounds and permeates all of the galaxies in our universe. This proposed dark matter makes up an estimated 80-95% of all mass in our universe. Dark matter has never been observed, and not a shred of proof has been found for its existence.

- Astronomers have discovered in recent years that our universe continues to expand. To explain this within the context of the Big Bang, the idea of "dark energy" is often promoted as the energy force that overcomes gravity and allows stars and galaxies to keep expanding. Like dark matter just mentioned, dark energy is completely hypothetical and has never been observed.

- The Big Bang proposes that an explosion originally started everything. The linear momentum of the objects that flew out from the explosion would have been in straight lines away from the explosion, because there would not have been any gravitational forces or objects in their way to change their paths. But astronomers and physicists point out that we observe angular momentum everywhere in the universe, as demonstrated when objects in space consistently travel in circular patterns and elliptical orbits. There is no way to scientifically explain how linear motions converted to consistently distinctive and patterned angular motions.

- Experimental physics has taught us that whenever energy produces matter, the chemical reaction always creates equal amounts of both matter and antimatter. If the Big Bang was true, our universe would be full of nearly equal amounts of matter and antimatter. What do we actually observe? Our universe consists of more than 99% matter and a very small amount of antimatter!

In 2004, 34 scientists from many disciplines wrote an open letter to the scientific community, stating that they no longer have faith in the Big Bang proposal. A reference to the proposal can be found in the list of resources for this reason. Since its original release, another 476 scientists, engineers, researchers, and additional signers from all over the world have added their names to the letter. The following set of statements from their letter identifies why the Big Bang fails to be valid science:

◀ *The big bang today relies on a growing number of hypothetical entities, things that we have never observed - inflation, dark matter and dark energy are the most prominent examples. Without them, there would be a fatal contradiction between the observations made by astronomers and the predictions of the big bang theory.*

But the big bang theory can't survive without these fudge factors. Without the hypothetical inflation field, the big bang does not predict the smooth, isotropic cosmic background radiation that is observed, because there would be no way for parts of the universe that are now more than a few degrees away in the sky to come to the same temperature and thus emit the same amount of microwave radiation.

In no other field of physics would this continual recourse to new hypothetical objects be accepted as a way of bridging the gap between theory and observation. It would, at the least, raise serious questions about the validity of the underlying theory.

What is more, the big bang theory can boast of no quantitative predictions that have subsequently been validated by observation. The successes claimed by the theory's supporters consist of its ability to retrospectively fit observations with a steadily increasing array of adjustable parameters. ▶ [1]

Astrophysicist and evolutionist Geoffrey Burbidge, Ph.D., writing in the journal *Scientific American*, helps us recognize that the Big Bang philosophy fails as real science:

◀ *Big Bang cosmology is probably as widely believed as has been any theory of the universe in the history of Western civilization. It rests, however, on many untested, and in some cases untestable, assumptions. Indeed, big bang cosmology has become a bandwagon of thought that reflects faith as much as objective truth.* ▶ [2]

A final thought on the Big Bang... The universe displays incredible beauty, patterns, order, complexity and design. Does it make any sense, scientific or not, to suggest that an explosion produced the beauty, the patterns, the order, the complexity, and the designs displayed in our universe? Has anyone, whether scientist or not, ever seen an explosion create beauty, patterns, order, complexity and design? The answer to both questions is a resounding, "Of course not!"

Scientific Reason 1 Summary: The suggestion of a Big Bang is not observable, cannot be scientifically substantiated, and merely offers hypothetical and philosophical ideas supporting evolution and its alleged billions of years of time.

⌘ Scientific Reason 2 ⌘
Laws of Planetary Motion

In the early 1600s, mathematician and astronomer Johannes Kepler discovered that planets in our solar system demonstrate three laws of planetary motion:

1. Planets orbit in ellipses, not perfect circles, with our sun as the focus of the ellipse.

2. Planets orbit our sun such that they sweep out equal areas in equal amounts of time. This also means they speed up slightly as they get closer to our sun, and they slow down slightly as they travel farther away from our sun.

3. Planets orbit around our sun in such a way that an exact mathematical relationship exists between the planet's distance from our sun and their orbital period (time to travel around the sun).

We also know that Kepler's laws apply to the orbits of moons around a planet within our solar system. Interestingly, mathematicians have learned that Kepler's laws of planetary motion can be derived mathematically from the laws of physics for gravity and motion, which Sir Isaac Newton discovered. This all stands in stark contrast against evolution's foundational premise that behaviors within our universe, including those of the stars and planets, come from random-chance processes mysteriously resulting from an unknown initial explosion billions of years ago (See Scientific Reason 1).

Scientific Reason 2 Summary: The planets in our solar system consistently follow precise orbits and patterns that can be described by Kepler's three laws of planetary motion. It is scientifically and logically impossible to conclude that random, purposeless, and unintelligent evolutionary processes are responsible for establishing the precise orbits and patterns displayed in our solar system.

⌘ Scientific Reason 3 ⌘ Spiral Galaxies

In our universe, an incredible number of beautiful galaxies display an overwhelming number of stars and planets. Astronomers estimate there are 100-200 billion galaxies populating our universe. Remarkably, astronomers have observed that an entire galaxy can orbit another entire galaxy without their individual stars crashing into one another! Astronomers estimate that an individual galaxy may contain 100 billion stars. Although stars sometimes appear spread throughout a galaxy, for many galaxies, the stars appear to be concentrated in areas known as spiral arms. We know this from the pictures taken by incredibly advanced technology such as the Hubble Telescope. These spiral arm galaxies often appear as a central nucleus region with arc-shaped arms of stars radiating outward. These arms are often described as "swirling" away from the galaxy nucleus, but no matter how we describe them, these galaxies and stars display incredible order and beauty!

Astronomers have discovered that the outermost stars in spiral arms take longer to complete their orbits around the galaxy's nucleus than the stars located in the innermost portion of the spiral arms. As galaxies spin, nothing is forcing the stars to stay within their respective spiral arms. Logically, after a few rotations, the spiral arms and their circling stars would disperse evenly throughout the galaxy. Note that astronomers have been able to view galaxies from all over our universe and consistently observe that many galaxies still contain spiral arms. If these galaxies really were billions of years old, as alleged by evolution and its long age premise, the spiral galaxies would have evenly dispersed by now.

Another important point to note: evolution proposes that the origin of galaxies was from a scientifically unknown and unsubstantiated explosion named the Big Bang. But it is scientifically and logically impossible that an explosion, which always causes randomness and chaos, initiated the precise order, complexity, and beauty observed in 100-200 billion galaxies and the 100 billion stars within each galaxy. In the same recognition, an explosion could never cause a massive galaxy with its 100 billion orbiting stars to be able to perfectly orbit another entire galaxy without crashing into each other!

Astronomer and evolutionist Martin Harwit shares the reality facing evolutionists who try to explain how our universe became filled with billions of stars in each galaxy:

◀ *The universe we see when we look out to its farthest horizons contains a hundred billion galaxies. Each of these galaxies contains another hundred billion stars. That's* $10^{(22)}$ *stars all told. The silent embarrassment of modern astrophysics is that we do not know how even a single one of these stars managed to form.* ▶ [3]

Scientific Reason 3 Summary: The existence of spiral galaxies throughout our universe contradicts the supposed billions of years of history. Due to the orbiting motions of individual galaxies and the stars within them, the spiral arms of galaxies would have already dispersed if the universe was billions of years old.

⌘ Scientific Reason 4 ⌘
The Poynting-Robertson Effect

In 1903, physicists John Henry Poynting and Howard Percy Robertson discovered that our sun has a gravitational force which causes small particles in space (also known as micro-meteoroids) to be pulled into our sun and disintegrated. An analogy often used to describe this observed behavior depicts our sun acting like a giant vacuum cleaner sweeping through space for these particles. We do observe that comets and asteroids are adding more space dust to our solar system as the comets and asteroids break apart and particles fall into unstable orbits. Astronomers have observed that increases in space dust occur at a rate which is less than half the rate that space dust is being vacuumed into our sun. The net impact of the Poynting-Robertson Effect reduces the total amount of space dust by pulling 100,000 tons of space dust into our sun every day! [5]

An explanation of how the Poynting-Robertson Effect works involves light radiating from our sun. Light is emitted as small, dual, wave-particles called photons. These photons fly away from the sun in straight lines in all directions. When a photon smacks into a small piece of space dust, the photon slows the speed of the space dust and changes its already unstable orbit. The space dust's reduction in speed and change in orbit means it can no longer escape the powerful gravitational force pulling it towards our sun, and thus the space dust becomes absorbed and disintegrated into our sun.

Three very interesting points arise because of the Poynting-Robertson Effect, all related to the age of our solar system and universe.

- First, by calculating and comparing how much space dust still exists in our solar system versus how quickly space dust is being vacuumed into our sun, we discover that the observed net rate of the Poynting-Robertson Effect means that our solar system cannot be more than 10,000 years old. [5]

- Second, astronomer Fred Whipple of Harvard University discovered that, over time, the Poynting-Robertson Effect would sort out and disperse meteor dust particles based on their mass. The Poynting-Robertson Effect causes the particles with less mass to be vacuumed into the sun faster than the particles with more mass because those with more mass are not slowed as quickly by the sun's photons. If our solar system was older than 10,000 years, we would observe distinctive patterns and segregated regions of space dust within our solar system, based upon the mass of the particles. However, no such distinctive space dust patterns or segregations are observed in our solar system. Astronomer and creationist Harold Slusher comments on this observation:

◀ *If there is not this dispersion, this spreading out of material into smaller and smaller orbits, how can meteor streams be more than 10,000 years old? Meteor showers do not show the dispersion effect that the Poynting-Robertson effect predicts; therefore, they cannot be old.* ▶ [4]

- Third, astronomers observe large amounts of space dust circling other stars in our universe. Because of the rate that the Poynting-Robertson Effect occurs, if our universe was really billions of years old, any space dust would have already been vacuumed and disintegrated into a nearby star. Because we still observe space dust circling other stars in our universe, this communicates a timeframe of thousands of years and not millions or billions of years. Dr. Walt Brown, who earned his Ph.D. in mechanical engineering at M.I.T. and served as an U.S. Air Force Colonel, sums up the significance of these observations:

◀ *Huge quantities of microscopic dust particles also have been discovered around some stars. Yet, according to the theory of stellar evolution, those stars are many millions of years old, so that dust should have been removed by stellar wind and the Poynting-Robertson effect. Until some process is discovered that continually resupplies vast amounts of dust, one should consider whether the 'millions of years' are imaginary.* ▶ [5]

Scientific Reason 4 Summary: Because of scientific observations of the Poynting-Robertson Effect on the space dust within our solar system as well as around other stars within our universe, we can deduce that our solar system cannot be more than 10,000 years old, which contradicts the beliefs of evolution and its claim of billions of years of history.

⌘ Scientific Reason 5 ⌘
Our Shrinking Sun

Astronomers have discovered our sun is shrinking. With a diameter of about 1 million miles, nobody notices this shrinking impact during the span of a normal lifetime. But this scientific observation has huge implications regarding the age of our earth.

More than one scientist has announced research and data that our sun is shrinking. In 1980, astronomer and evolutionist John A. Eddy, Ph.D., from the Smithsonian

Center for Astrophysics and High Altitude Observatory in Boulder, CO, and mathematician and evolutionist Aram Boornazian found evidence that the sun is shrinking about 0.1% per century. This rate equates to about 5 feet per hour or 1 percent every 1,000 years. Based on his research, Dr. Eddy states:

◀ *11,210,959 years ago the Sun would have been touching the Earth.* ▶ [6]

Based on his research of published works that had been released between 1715 and 1979, astronomer and evolutionist David W. Dunham states in a *Science* journal article:

◀ *The Sun would have been touching the Earth about 158,272,358 years ago.* ▶ [7]

Obviously both astronomers are providing estimates on how big they think the sun would have been if it had existed millions of years ago. Physicist and creationist Russell Akridge, Ph.D., highlights the implications of such scientific observations:

◀ *Over 100,000 years these changes would have accumulated so much that life of any kind on the earth would have been very difficult, if not impossible. Thus, all life on the earth must be less than 100,000 years old. The sun, 20 million years ago, would have been so large that it would have engulfed the earth. The earth cannot be more than 20 million years old. Those dates as upper limits rule out any possibility of evolution requiring hundreds of millions of years... The changes detected in the sun call into question the accepted thermonuclear fusion energy source for the sun. This, in turn, questions the entire theoretical structure upon which the evolutionary theory of astrophysics is built.* ▶ [8]

Note the significance of these scientific observations... Because our sun is shrinking, we can recognize that even only an alleged 100,000 years ago the sun would have produced too much power, heat, and energy for life to exist on the earth. This scientific research certainly does not match with evolution's premise of billions of years of earth's history.

Scientific Reason 5 Summary: The observed rate of our shrinking sun means that our sun and our earth cannot be millions or billions of years old, contrary to the beliefs essential to evolution and its long age premise.

⌘ Scientific Reason 6 ⌘
Heat Radiation of Jovian Planets

Astronomers have discovered that Jupiter, Saturn, Uranus, and Neptune, together known as the Jovian planets, each radiate more heat than they absorb from our sun.

Because Jupiter, Saturn, Uranus, and Neptune all give off more heat than they absorb from our sun, it is logical to conclude they each have an internal heat energy source. Both Jupiter and Saturn give off about twice the amount of energy that they absorb from our sun. Physicist and creationist Ron Samec, Ph.D., highlights this:

◀ *Jupiter, Saturn, Uranus and Neptune all give off more heat than they receive. The power excess for Jupiter is 3 x* $10^{(17)}$ *watts. Jupiter actually radiates, mostly in the infrared, nearly twice as much power as it receives from the Sun. Saturn, although generating about half as much total power as Jupiter, has only 30% of its mass. So it produces nearly twice the energy per unit mass than Jupiter! Uranus and Neptune also produce excess energy but much less in total than their big brothers.* ▶ [9]

Logically, if our Jovian planets were really billions of years old, they would have cooled off by now and would not be radiating more energy than they are absorbing from our sun.

Scientific Reason 6 Summary: The Jovian planets in our solar system radiate more energy than they are absorbing from our sun, which is opposite to what is expected if they were billions of years old. This scientific observation of our Jovian planets directly contradicts evolution's premise of billions of years.

⌘ Scientific Reason 7 ⌘
Short-Lived Comets

Short-lived comets are chunks of rock, dust, and ice in our solar system which orbit our sun. Each time they pass by our sun, the sun's solar wind and solar radiation breaks away part of the short-lived comet. This stream of materials blasting away creates the tail for which comets are known. Considering the amount of ice and debris breaking away from comets each time they pass by our sun, these comets can only survive a limited number of orbits. Astronomers have discovered that short-lived comets in our solar system usually have a lifespan of 200 years or less. Scientific observations of comets and their orbits within our solar system have identified that the ages of our short-lived comets do not harmonize with evolutionary timescales of billions of years. [10]

In order to support the idea that comets in our solar system could be billions of years old, some astronomers have suggested that an Oort Cloud or a Kuiper Belt is constantly releasing new comets. But after years of searching, the existence of neither an Oort Cloud nor a Kuiper Belt has been conclusively confirmed. Evolutionists Carl Sagan and Ann Druyan highlight that the proposed Oort Cloud has never been observed:

◀ *Many scientific papers are written each year about the Oort Cloud, its properties, its origin, its evolution. Yet there is not yet a shred of direct observational evidence for its existence.* ▶ [11]

Scientific Reason 7 Summary: The existence of short-lived comets in our solar system, whose ages do not harmonize with evolutionary timescales of billions of years and for which, as of today, no conclusive evidence exists that they are being replenished from anywhere in our universe, contradicts evolution and its alleged billions of years of history.

⌘ Scientific Reason 8 ⌘ Supernovas

Astronomers classify a star that violently explodes as a supernova. After the explosion, a massive continually-expanding cloud of debris flies out from the explosion. This expanding cloud of debris is known as a Supernova Remnant (SNR).

One of the most well-known examples of a SNR was formed in 1054 A.D. Today we call it the Crab Nebula. It was so large, powerful, and bright when it formed that it was even visible to the naked eye during the daytime!

Astronomers note that a supernova briefly outshines much of the remainder of its galaxy, and then consistent with the laws of physics, the resulting SNR continues to expand. According to physics models used to calculate how fast a SNR expands, a SNR will grow to a diameter of 300 light years. Remember that a light year is not a measurement of time but rather a measurement of distance; it is the distance that light at its current speed will travel for 1 year in a vacuum (void of any forces to slow its speed). Thus, a light year distance is roughly 5.9 million million miles. Astronomers recognize that considering both their massive size and the modern advances in space viewing technologies, we should be able to see many SNRs within our universe.

Astronomers have created a system to categorize the three stages of SNRs. Chemist and creationist Jonathan Sarfati, Ph.D., identifies that based upon definitions of these three stages of SNRs, if our Milky Way Galaxy was actually billions of years old, we would observe two stage one SNRs, about 2,260 stage two SNRs, and about 5,000 stage three SNRs. So what do we actually observe? Astronomers observe two stage one SNRs as expected, but observe only 200 stage two SNRs and no stage three SNRs. These observations are actually expected for a universe with an age of less than 10,000 years! [12]

Astrophysicist and creationist David Russell Humphreys, Ph.D., summarizes the scientific reality of these observations:

◀ *According to astronomical observations, galaxies like our own experience about one supernova (a violently-exploding star) every 25 years. The gas and dust remnants from such explosions (like the Crab Nebula) expand outward rapidly and should remain visible for over a million years. Yet the nearby parts of our galaxy in which we could observe such gas and dust shells contain only about 200 supernova remnants. That number is consistent with only about 7,000 years' worth of supernovas.* ▶ [12]

Our Milky Way Galaxy has a sister galaxy named the Large Magellanic Cloud Galaxy. Rather than displaying the number of SNRs expected for a galaxy billions of years old, it too only displays a number of SNRs that are consistent with an age of 10,000 or fewer years. [13]

Scientific Reason 8 Summary: The number of SNRs that should be observed if our galaxy was actually billions of years old do not exist, but rather the number of observed SNRs in our galaxy points to an age of 10,000 or less years. This lack of observed SNRs scientifically contradicts the evolutionary assumption of billions of years.

✍ Chapter 1 Endnotes ✍

1 – "An Open Letter to the Scientific Community", *New Scientist*, Cosmologystatement.org (22May2004). Web 15September2015. <https://web.archive.org/web/20140401081546/http://cosmologystatement.org>

2 – Geoffrey Burbidge, "Why Only One Big Bang?", *Scientific American* (February 1992) page 120.

3 – Martin Harwit, "Book Reviews", *Science* (March 1986) pages 1,201-1,202.

4 – Harold S. Slusher, "A Young Universe", *Bible-Science Newsletter* (13January1975).

5 – Walt Brown, *In the Beginning: Compelling Evidence for Creation and the Flood* (Phoenix, AZ: Center for Scientific Creation, 1995) page 40.

6 – John A. Eddy and Aram A. Boornazian, "Secular Decrease in the Solar Diameter, 1863-1953", *Bulletin of the American Astronomical Society*, Vol. 11, No. 2 (1979) page 437.

7 – David W. Dunham, "Observations of a Probable Change in the Solar Radius between 1715 and 1979", *Science*, Vol. 210 (12December1980) pages 1,243-1,245.

8 – Russell Akridge, "The Sun Is Shrinking", *Acts & Facts*, Vol. 4 (1980) page 9. <http://www.icr.org/article/sun-shrinking>

9 – Ron Samec, "The Age of the Jovian Planets", *Journal of Creation,* Vol. 14, Iss. 1 (April 2000) pages 3-4. <http://creation.com/the-age-of-the-jovian-planets>

10 – Wayne Spencer, "Our young solar system", *Creation,* Vol. 33, Iss. 4 (October 2011) pages 38-41. <http://creation.com/young-solar-system>

11 – Carl Sagan and Ann Druyan, *Comets* (New York: Random House, 1985) page 201.

12 – Jonathan Sarfati, "Exploding stars point to a young universe", *Creation,* Vol. 19, Iss. 3 (June 1997) pages 46-48. <http://creation.com/exploding-stars-point-to-a-young-universe>

13 – David Russell Humphreys, "Evidence for a Young World", *Acts & Facts*, Vol. 6 (2005) page 34. <http://www.icr.org/article/evidence-for-young-world>

Chapter 1 Resources

Big Bang

- https://web.archive.org/web/20140401081546/http://cosmologystatement.org (Open Letter to Scientific Community)
- https://answersingenesis.org/big-bang/the-big-bang
- https://answersingenesis.org/big-bang/what-about-the-big-bang/#star
- http://creation.com/astronomy-and-astrophysics-questions-and-answers#big_bang
- https://answersingenesis.org/astronomy/age-of-the-universe/evidence-for-a-young-world
- http://creation.com/did-god-use-big-bang
- http://creation.com/the-mind-of-god-and-the-big-bang
- http://creation.com/secular-scientists-blast-the-big-bang
- http://creation.com/big-bang-not-a-reason
- *The New Answers Book 2: Chapter 10* by Answers in Genesis
- *Creation Basics and Beyond: Chapter 44* by Institute For Creation Research

Laws of Planetary Motion

- http://creation.com/johannes-kepler
- http://www.icr.org/fb/Kepler

Spiral Galaxies

- https://answersingenesis.org/astronomy/cosmology/galaxies-unexplained-spirals
- http://creation.com/early-galaxies-don-t-fit
- http://www.icr.org/creation-galaxy

The Poynting-Robertson Effect

- http://planetfacts.org/poynting-robertson-effect
- http://www.creationscience.com/onlinebook/AstroPhysicalSciences52.html
- http://creationism.org/ackerman/AckermanYoungWorldChap03.htm

Our Shrinking Sun

- http://www.answersingenesis.org/articles/tj/v15/n2/faint-sun
- http://www.icr.org/article/sun-shrinking
- http://creation.com/the-sun-our-special-star

Heat Radiation of Jovian Planets

- http://www.icr.org/article/solar-system-neptune
- http://creation.com/the-energy-balance-of-uranus-implications-for-special-creation
- http://creation.com/jupiter-king-of-the-planets-and-testament-to-our-creator
- http://creation.com/saturn-the-ringed-planet
- http://creation.com/the-age-of-the-jovian-planets
- https://answersingenesis.org/astronomy/solar-system/uranus
- https://answersingenesis.org/astronomy/solar-system/neptune-monument-to-creation
- https://answersingenesis.org/astronomy/solar-system/anomalies-planets-and-satellites-solar-system-indication-design

Short-Lived Comets

- http://creation.com/young-solar-system
- http://creation.com/comets-portents-of-doom-or-indicators-of-youth
- http://creation.com/comets-and-the-age-of-the-solar-system
- http://www.answersingenesis.org/articles/tj/v11/n3/comets
- http://www.answersingenesis.org/articles/tj/v16/n2/comets
- http://www.answersingenesis.org/articles/tj/v15/n2/oort

Supernovas

- http://creation.com/exploding-stars-point-to-a-young-universe
- http://www.icr.org/article/evidence-for-young-world
- http://www.icr.org/article/6373/245
- https://answersingenesis.org/astronomy

⌘ Chapter 2 ⌘

Physical Science

What is a big deal - the biggest deal of all - is how you get something out of nothing. Don't let the cosmologists try to kid you on this one. They have not got a clue either - despite the fact that they are doing a pretty good job of convincing themselves and others that this is really not a problem. 'In the beginning,' they will say, 'there was nothing - no time, space, matter or energy. Then there was a quantum fluctuation from which...' Whoa! Stop right there. You see what I mean? First there is nothing, then there is something. [1]

⌘ Scientific Reason 9⌘
Matter

We all recognize that the physical world is arrayed in the plants, animals, and scenic landscapes around us. Maybe we hike in the mountains, or grow gardens in soil, or build things out of wood or metal, but we very rarely stop to think and ask ourselves where matter and all the material things around us originated.

The law of cause and effect demands that anything that exists must have a cause - there is no exception to this law and it has never been shown to be false. The predominant belief within evolution is that all matter somehow mysteriously appeared via a massive explosion known as the Big Bang. The Big Bang suggests that an incredibly compacted "ball" known as a singularity exploded, expanded, and led to all the matter we see all around us. (See Scientific Reason 1) But logically, if matter exists, it had to have a source (a cause). To say that matter originated via a Big Bang still does not explain where the matter originally came from (what caused it to exist), and evolution has no answer for where all matter originated. Astronomer and science writer David Darling makes a powerful observation noting that evolution cannot account for the existence of matter:

◀ *What is a big deal - the biggest deal of all - is how you get something out of nothing. Don't let the cosmologists try to kid you on this one. They have not got a clue either - despite the fact that they are doing a pretty good job of convincing themselves and others that this is really not a problem. 'In the beginning,' they will say, 'there was nothing - no time, space, matter or energy. Then there was a quantum fluctuation from which...' Whoa! Stop right there. You see what I mean? First there is nothing, then there is something. And the cosmologists try to bridge the two with a quantum flutter, a tremor of uncertainty that sparks it all off. Then they are away and before you know it, they have pulled a hundred billion galaxies out of their quantum hats... You cannot fudge this by appealing to quantum mechanics. Either there is nothing to begin with, in which case there is no quantum vacuum, no pre-geometric dust, no time in which anything can happen, no physical laws that can effect a change from nothingness into somethingness; or there is something, in which case that needs explaining.* ▶ [1]

Scientific Reason 9 Summary: Anything that exists must have a cause, and according to the rules of logic, it must therefore have come from a source. Evolution cannot scientifically explain the source or cause of anything, including all matter.

⌘ Scientific Reason 10 ⌘
Atoms and Elements

Atoms are the basic building blocks of all matter and physical things. Here are seven interesting attributes of atoms:

- Atoms are the smallest particle that retains the properties of a specific element. Atoms consist of three parts: protons, neutrons, and electrons. To put their incredibly small size into perspective, an average piece of human hair measures about 100 nanometers in diameter, and a million carbon atoms could be lined up next to one another within the width of this piece of hair.

- The three parts of an atom function like a miniature solar system, with electrons swirling around the nucleus in layers called shells, while the neutrons and protons reside in the nucleus. These shells each allow only a specific number of electrons to swirl within them.

- The negative electromagnetic charge of electrons keeps them in orbit around the positively charged nucleus, where positively charged protons and neutrally charged neutrons reside. It is estimated that the electromagnetic force of an atom's nucleus (which pulls on the swirling electrons) is about 1,000 times stronger than the force of gravity we experience from the earth. To put this into perspective, we would not be able to lift our foot off the ground if the earth's force of gravity on us was as strong as the forces within an atom!

- The atom's nucleus is only about 1 ten-thousandth of the size of the entire atom yet still contains the majority of the atom's mass, because protons and neutrons have 1,800 times more mass than an electron.

- The number of protons determine the type of atom. For example, a normal oxygen atom has eight protons. It is estimated that there are about $10^{(20)}$ atoms of oxygen in just one drop of water.

- Atoms have an incredible ability to interact with other atoms in order to create compounds. For example, chemists estimate about 10 million compounds can be made with a carbon atom. A common example of a carbon-based compound is the carbon dioxide we exhale.

- Made with just one electron and one proton, normal hydrogen atoms are the lightest, simplest, and most common atom in our universe, estimated to make up over 70% of the mass of our universe.

Atoms are the smallest unit of an element such as carbon. A single atom possesses the characteristics of its element. The properties of all the elements are not arbitrary; rather they are incredibly patterned and distinctively organized. They are so organized and orderly that we have a Periodic Table of the Elements. Here are eight significant characteristics about elements and the Periodic Table:

- The Periodic Table is named after its rows, which are called periods.

- As of 2010, there are 118 elements in the Periodic Table; 98 are naturally occurring and the other 20 have been developed by scientists altering one of the naturally occurring elements. The man-made elements sometimes only exist for a fraction of a second.

- The columns of the Periodic Table are sets of groups. The elements in each of the 18 groups all have the same number of outermost electrons, which are known as valence electrons. In general, elements on the left side of the Periodic Table are metals and those on the right side are non-metals. In the left-most group within the Periodic Table, elements have one electron in their outer shell and elements in the right-most group have a full outer shell of eight electrons.

- The group four elements will bond with just about any element that is capable of bonding with another element. A notable example is silicon. Silicon atoms readily bond with other silicon atoms to form the crystalline structures used to make semiconductors.

- The right-most column of the Periodic Table lists the noble gases, which have closed electron shells, meaning they normally do not take any outside additional electrons into their atoms. The left-most column identifies the alkali metals, which have only one outer shell electron and are highly reactive (meaning they bond easily with other atoms). The second through 17th columns each have their own specific properties as well.

- The Periodic Table provides the Atomic Mass and the Atomic Number of each element. The Atomic Number is the number of protons in the nucleus, and it impacts the number of electrons swirling around the nucleus and thus the chemical properties of the atom. Modern day versions of the Periodic Table usually sort the elements from left to right by their Atomic Number.

- There are behavior trends for the elements which line up in diagonal patterns across the chart, such as the metallic character of the elements or the boiling points of the elements.

The Periodic Table of the Elements reflects incredible order, structure, patterns, and consistent behaviors of the elements. It is not scientifically reasonable that an explosion and random-chance evolutionary processes are responsible for the origin, complexity, beauty, design, and consistent behaviors displayed by the elements.

Scientific Reason 10 Summary: Atoms are an orderly reality of our physical existence that demonstrate remarkable structure, consistency, stability, precision, and interconnectivity, none of which can be the result of the random-chance processes which evolution offers. The design, structure, and behavior of elements, which are beautifully and orderly arranged in the Periodic Table of Elements, cannot be the result of evolution and its unintelligent, random, and goalless processes.

⌘ Scientific Reason 11 ⌘ Water

It goes without saying that water is incredible. Water is absolutely necessary for life. It makes up an estimated 60-70% of our bodies, and it contributes to our daily lives in fascinating ways. Here are six unique life-supporting attributes of water that make it unequaled in comparison to any other substance.

- Our earth benefits from a water cycle, in which molecules of water rotate between the oceans and the atmosphere. Water moves to the atmosphere by the process of evaporation and back to the oceans through rain, snow, and water flowing down our rivers. We seldom think about this, but any system such as our water cycle demonstrates the principle of irreducible complexity. In order for the water cycle to happen, all the parts of the water cycle have to fully exist and fully work. For example, it would be impossible for the water cycle to only involve rain and snow, because without evaporation water would not be able to return to the atmosphere and be available as rain or snow again. Evolution is based upon the belief that small, successive physical changes over billions of years caused everything to exist, but there is no way for the water cycle to occur with only partial rain, partial snow or partial evaporation processes in place. In reality, in order to exist, all the parts of the water cycle within the earth's ecosystem must exist and be fully functional.

- Ice normally floats on top of liquid water. If this was not so, many forms of life could not exist on the earth. Why? Imagine what would happen if a lake froze first from the bottom and then progressively up to the top; nearly all of the living creatures in the lake would die.

- Water has an attribute called a specific heat value, which is a measurement of how fast or slow water heats up and stores energy. The enormous specific heat capacity of water means that our oceans heat up slowly, and inversely, they cool down slowly. This attribute of water helps protects our earth from sudden and dramatic climate changes.

- Water demonstrates a property called surface tension, observed when the surface of a body of water acts like an elastic sheet between the water and the air above. A water strider gliding across the surface of a pond without sinking provides an easy-to-see benefit of surface tension.
- Water displays cohesion, observed when water molecules "stick" to one another, so to speak. Water's cohesive property is instrumental in allowing trees and plants to bring water up from their roots against the constant force of gravity pulling down on the water inside their roots, trunks, and branches.
- Water's capacity to act as a solvent gives it the ability to dissolve other substances, such as sugars and salts. Without this capability to break down other substances, our stomach and digestive systems would not work.

Scientific Reason 11 Summary: All of the unique, finely calibrated, life-giving attributes of water are impossible to scientifically explain as the result of evolution's random and unintelligent processes.

⌘ Scientific Reason 12 ⌘ Universal Constants

Universal constants are precise mathematical and scientific values which are instrumental to the Laws of Physics. These universal constants represent significant physical realities, from the strength of gravitational forces pulling on everything that has mass to the behavior of electrons swirling within every atom.

Scientists recognize that more than 40 universal constants play a significant role in our physical reality, although not all scientists agree on the exact list. Here are some universal constants which are commonly recognized:

- The speed that light travels when there is nothing to inhibit its path
- The strength of electrical currents
- The strength of magnetism
- The pull of gravity by one object upon another object
- The amount of mass in an atom's electrons and protons

The point here is not to dive into the specifics and complexities of each universal constant but rather to highlight that these universal constants must exist at specific values in order for life to exist. If they were not finely and precisely tuned, or they failed to behave in a consistent way, or they were altered even a very slight amount, life on earth would not be possible! As an example, at the most fundamental level of all material things (atoms), if the electrical, gravitational, or mass constants of protons, neutrons, or electrons in an atom were altered even a fraction of one percent, the physical world would not exist.

Astronomer and mathematician Sir Fred Hoyle, Ph.D., recognizes that the properties of atoms and universal constants could never originate from the random chance-based belief of evolution:

◀ *Would you not say to yourself, some super-calculating intellect must have designed the properties of the carbon atom, otherwise the chance of my finding such an atom through the blind forces of nature would be utterly minuscule. A common sense interpretation of the facts suggests that a superintellect has monkeyed with physics, as well as with chemistry and biology, and that there are no blind forces worth speaking about in nature.* ▶ [2]

Physicist and evolutionist Paul Charles William Davies, Ph.D., in his book *The Accidental Universe*, notes the precise distinction and values of the universal constants:

◀ *The numerical values that nature has assigned to the fundamental constants such as the charge of the electron, the mass of the proton, and the Newtonian gravitational constant, may be mysterious, but they are critically relevant to the structure of the universe that we perceive... Had nature opted for a slightly different set of numbers, the world would be a very different place. Probably we would not be here to see it.* ▶ [3]

The significant point here, as emphasized by Dr. Hoyle and Dr. Davies, is that if even one of these universal constants was changed even a fraction of one percent, our universe as we know it would not exist nor be able to support life on earth.

Scientific Reason 12 Summary: It is not scientifically reasonable to suggest that random and purpose-less processes promoted by evolution could have precisely and accurately calibrated all of the universal constants to the specific and exact values required for life to exist on earth.

⌘ Scientific Reason 13 ⌘
Laws of Physics

Laws of physics describe how our universe behaves. Many of these laws are mathematical in nature and follow precise formulas. Here are a few example laws of physics:

- Newton's Laws of Motion
- Newton's Law of Gravity
- The Conservation of Mass and Energy
- The Conservation of Momentum
- The Laws of Thermodynamics
- Electrostatic Laws
- The Speed of Light

Entire books have been written about each of these physical laws, and the point here is not to focus on the specific details of each law. Rather, we need to recognize that life is only possible because material things behave in specific and consistent behaviors that can be described by the laws of physics. If these laws of physics behaved inconsistently at different places on the earth or out in space, life could not exist. For example, if the force of gravity altered from very weak to very strong at different places on the earth, life would be impossible. In such a hypothetical scenario, for example, the following animal behaviors would be harmed by the shifting and fluctuating gravitational forces: migration, species dispersion, bird and insect flight, and the ability for animals to relocate. All scientists know that the consistent behavior of these laws of physics is critical for life to exist.

M.I.T astrophysicist and evolutionist Alan Guth, Ph.D., provides a candid answer that evolution cannot scientifically explain the origin of the laws of physics:

◀ *Where do the laws of physics come from? We are a long way from being able to answer that one.* ▶ [4]

Scientific Reason 13 Summary: It is not scientifically reasonable to believe that random, designless, chance-based processes promoted by evolution precisely established and continually maintain the consistent laws of physics required for life on earth to exist.

⌘ Scientific Reason 14 ⌘
Second Law of Thermodynamics

The Second Law of Thermodynamics describes the way energy behaves. In summary, the Second Law of Thermodynamics identifies that all energy forms and material things are moving from a state of higher organization to a state of lower organization. There are no known exceptions to this law - we observe its consistent behavior everywhere in our universe.

The result of the Second Law of Thermodynamics means that we exist in an environment where order moves to disorder, and we observe over time an increase in entropy (material things breaking down or degenerating), greater randomness, greater disorganization, and greater decay of energy and matter in our universe. But evolution requires an ongoing increase in order, a constant gain of new information, and continual advancements in the structure and design of material things, all of which are impossible according to the Second Law of Thermodynamics. This truth about our universe is not a debated issue, as biochemist and evolutionist Isaac Asimov, Ph.D., explains in simple terms:

◀ *Another way of stating the Second Law, then, is: 'The universe is constantly getting more disorderly.' Viewed that way, we can see the Second Law all about us. We have to work hard to straighten a room, but left to itself, it becomes a mess again very quickly and very easily. Even if we never enter it, it becomes dusty and musty. How difficult to maintain houses, and machinery, and our own bodies in perfect working order; how easy to let them deteriorate. In fact, all we have to do is nothing, and everything deteriorates, collapses, breaks down, wears out, all by itself - and that is what the Second Law is all about.* ▶ [5]

Many scientists promote the belief in evolution over billions of years in spite of the truth of the Second Law of Thermodynamics, and only a few are willing to recognize the inconsistent and unscientific thinking going on here. To his credit, sociologist and evolutionist Jeremy Rifkin recognizes this:

◀ *The Entropy Law says that evolution dissipates the overall available energy for life on this planet. Our concept of evolution is the exact opposite. We believe that evolution somehow magically creates greater overall value and order on earth. Now that the environment we live in is becoming so dissipated and disordered that it is apparent to the naked eye, we are for the first time beginning to have second thoughts about our views on evolution, progress, and the creation of things of material value... There will also be those who will stubbornly refuse to accept the fact that the Entropy Law reigns supreme over all physical reality in the world. They will insist that the entropy process only applies in selective instances and that any attempt to apply it more broadly to society is to engage in the use of metaphor. Quite simply, they are wrong.* ▶ [6]

Scientific Reason 14 Summary: It is not scientifically and logically reasonable to believe that evolution's random-chance processes are continually causing an increase in the order, structure, and complexity of living things, because this completely contradicts the Second Law of Thermodynamics.

⌘ Scientific Reason 15 ⌘
Formation of Stalactites, Diamonds, and Coal

One of the arguments used to support evolution is that the earth and everything in the universe is billions of years old. As supposed proof of these billions of years, the formation of stalactites, diamonds, and coal are assumed to have occurred over millions of years. Actually, we now know that stalactites can form in only decades, and diamonds and coal can form in less than a year.

We have examples of stalactites growing over three-eighths of an inch in one year, which is equivalent to a growth rate of over three feet in one century. For example, in the geological Museum of Stone located in the Altai Republic of Siberia, stalactite rings have growth rates of an inch every couple years. [7] In the Sequoyah Caverns in Alabama, stalactites have been observed to grow 10 inches in 10 years. [8] And at the Lincoln Memorial in Washington, D.C., stalactites beneath the memorial have grown over five feet since the memorial was built in 1923. These are just three examples of scientific observations showing us that stalactites form much more quickly than evolutionary timetables suggest. [8]

Diamond manufacturers can now successfully create diamonds in a relatively short period of time. A process developed in the 1980s called chemical vapor deposition and another process using high temperatures and pressure are both current diamond producing methods. A company named Pure Grown Diamonds in New York creates diamonds in a matter of days. [9]

Similarly, we have learned that coal can be created in less than a year with the right combination of extreme heat and pressure. [10]

Scientific Reason 15 Summary: Evolution relies on the idea that our earth is billions of years old, and the formation of stalactites, diamonds, and coal are commonly promoted as supporting geologic processes that require millions of years. Because we now observe that stalactites can be formed in decades and diamonds and coal in less than a year, we now have scientific support that millions of years are not needed.

⌘ Scientific Reason 16 ⌘ Formation of Granite

Evolution promotes the idea that granite, the dominant rock type in many regions on the earth, took millions of years to form. Recent scientific research has shown that granite has formed rapidly and did not require millions of years.

Geologist and creationist John Woodmorappe, M.A., identifies that granite can actually form quickly rather than needing millions of years as proposed by evolution:

◀ *The viscosities of silica-rich granitic magmas are orders of magnitude less than has been conventionally believed for over a hundred years. Consequently, the millions of years, previously believed necessary to form granitic rocks, are no longer required. The underlying physical processes involved in the segregation, transport and emplacement of granitic magmas operate on a timescale of months to centuries. Besides magma viscosity, important factors controlling the rate of emplacement include tectonic deformation of the crust, the protolith structure after partial melting, and the emplacement of magma by dike networks. Granitic magmatism is a rapid, dynamic process.* ▶ [11]

Geologist and creationist Andrew Snelling, Ph.D., also identifies that granite can form in a matter of days or years:

◀ *Even the conventional long-ages geologic community now regards the formation stages of granite plutons, after partial melting of source rocks to form granitic melts, that is, melt segregation, ascent and emplacement, to be 'geologically extremely rapid - perhaps even catastrophic.'* ▶ [12]

Physicist and creationist Robert Gentry, Ph.D., researched the rate of granite formation, and discovered that granite found all over the earth formed very rapidly. His discoveries include incredible research into Polonium halos existing within granite found all over the earth. These Polonium halos are colored rings left behind within granite when radioactive Polonium decays, and these halos only exist for an extremely short amount of time (seconds), thus providing support that the granite rock all over the world was formed very rapidly. More about Dr. Gentry's research and his fascinating insights into these granite halos can be found on his website titled *Halos* listed in the resources for this reason. From Dr. Gentry's site, here is a summary of his research:

◀ *Etched within Earth's foundation rocks - the granites - are beautiful microspheres of coloration, halos, produced by the radioactive decay of primordial polonium, which is known to have only a fleeting existence. The following simple analogy will show how these polonium microspheres - or halos - contradict the evolutionary belief that granites formed as hot magma slowly cooled over millions of years. To the contrary, this analogy demonstrates how these halos provide unambiguous evidence of both an almost instantaneous creation of granites and the young age of the earth.*

A speck of polonium in molten rock can be compared to an Alka-Seltzer dropped into a glass of water. The beginning of effervescence is equated to the moment that polonium atoms began to emit radioactive particles. In molten rock the traces of those radioactive particles would disappear as quickly as the Alka-Seltzer bubbles in water. But if the water were instantly frozen, the bubbles would be preserved. Likewise, polonium halos could have formed only if the rapidly 'effervescing' specks of polonium had been instantly encased in solid rock.

An exceedingly large number of polonium halos are embedded in granites around the world. Just as frozen Alka-Seltzer bubbles would be clear evidence of the quick-freezing of the water, so are these many polonium halos undeniable evidence that a sea of primordial matter quickly 'froze' into solid granite. The occurrence of these polonium halos, then, distinctly implies that our earth was formed in a very short time. ▸ [13]

Scientific Reason 16 Summary: We now have scientific research detailing that granite from all over the world formed in a very short timeframe, which contradicts evolution and its reliance upon billions of years of earth's history.

Chapter 2 Endnotes

1 – David Darling, “On Creating Something from Nothing”, *New Scientist*, Vol. 151 (14September1996) page 49.

2 – Fred Hoyle, “The Universe: Past and Present Reflections”, *Engineering and Science* (November 1981) pages 8–12.

3 – Paul Charles William Davies, *The Accidental Universe* (Cambridge: Cambridge University Press, 1982) Preface.

4 – Brad Lemley, “Guth's Grand Guess”, *Discover*, Vol. 23 (April 2002) page 38.

5 – Isaac Asimov, “In the Game of Energy and Thermodynamics You Can’t Even Break Even”, *Smithsonian Institute Journal* (June 1970) pages 10-11.

6 – Jeremy Rifkin, *Entropy: A New World View* (New York: Viking Press, 1980) pages 8 and 55.

7 – David Lewis, “Rapid Stalactite Growth in Siberia”, *Creation*, Vol. 32, Iss. 1 (January 2010) pages 40-42. <http://creation.com/rapid-stalactite-growth>

8 – Stephen Meyers and Robert Doolan, “Rapid stalactites”, *Creation*, Vol. 9, Iss. 4 (September 1987) pages 6-8. <http://creation.com/rapid-stalactites>

9 – Richard Fangrad, “Diamonds in days (actually, minutes!)”, Creation.com, Pub 15April2008. Web 18September2015. <http://creation.com/diamonds-in-days-actually-minutes>

10 – Stuart E. Nevins, “The Origin of Coal”, *Acts & Facts*, Vol. 11 (1976) page 5. <http://www.icr.org/article/origin-coal>

11 – John Woodmorappe, “The Rapid Formation of Granitic Rocks: More Evidence”, *Journal of Creation*, Vol. 15, No. 2 (August 2001) pages 122-125. <https://answersingenesis.org/geology/rocks-and-minerals/the-rapid-formation-of-granitic-rocks-more-evidence>

12 – Andrew A. Snelling, “Catastrophic Granite Formation”, *Answers Research Journal*, Pub 6February2008. Web 18September2015. <https://answersingenesis.org/geology/catastrophism/catastrophic-granite-formation>

13 – Robert Gentry, *Creation's Tiny Mystery* (Knoxville, TN: Earth Science Association, 1988). <http://www.halos.com>

Chapter 2 Resources

Matter

- http://creation.com/models-of-the-origin-of-the-universe

Atoms and Elements

- http://creation.com/atoms-and-gods-order
- http://creation.com/the-elements-of-the-universe-point-to-creation
- http://creation.com/blowing-old-earth-belief-away-helium

Water

- http://creation.com/the-wonders-of-water
- https://answersingenesis.org/blogs/danny-faulkner/2015/01/06/water-water-everywhere-but-not-a-drop-on-earth

Universal Constants

- http://creation.com/the-universe-is-finely-tuned-for-life

Laws of Physics

- https://answersingenesis.org/physics
- https://answersingenesis.org/is-god-real/god-natural-law
- *Creation Basics and Beyond: Chapter 46* by Institute For Creation Research

Second Law of Thermodynamics

- http://creation.com/thermodynamics-and-order-questions-and-answers
- http://creation.com/the-second-law-of-thermodynamics-answers-to-critics
- https://answersingenesis.org/physics/second-law-of-thermodynamics

Formation of Stalactites, Diamonds, and Coal

- http://creation.com/rapid-stalactites
- http://creation.com/rapid-stalactite-growth
- http://creation.com/diamonds-in-days-actually-minutes
- http://www.icr.org/article/origin-coal
- http://creationwiki.org/Diamond
- *Evolution's Achilles Heel* by Creation Ministries International

Formation of Granite

- http://www.halos.com (Dr. Gentry's Research)
- http://creation.com/the-rapid-formation-of-granitic-rocks-more-evidence
- http://creation.com/granite-formation-catastrophic-in-its-suddenness
- https://answersingenesis.org/age-of-the-earth/its-about-time
- https://answersingenesis.org/geology/rocks-and-minerals/the-rapid-formation-of-granitic-rocks-more-evidence
- https://answersingenesis.org/geology/catastrophism/catastrophic-granite-formation
- https://answersingenesis.org/geology/catastrophism/rapid-granite-formation
- https://answersingenesis.org/age-of-the-earth/raising-the-bar-on-creation-research

⌘ Chapter 3 ⌘

Earth

It is hereby submitted that most scientists are guilty of an overly-zealous interpretation of the doctrine of uniformitarianism. Many instructors dismiss the possibilities of global catastrophes altogether, whereas others ridicule and scoff at the early ideas. These same instructors will implore their students to think scientifically and to develop the principles of multiple-working hypotheses. The fact is, the doctrine of uniformitarianism is no more 'proved' than some of the early ideas of world-wide cataclysms have been disproved. [9]

⌘ Scientific Reason 17 ⌘
Anthropic Principle

The Anthropic Principle has been coined the "law of human existence". In summary, this principle recognizes there are many specifically calibrated attributes uniquely true about our earth, which distinctively enable it exclusively to support life. While not exhaustive, here are 15 attributes which contribute to the earth's Anthropic Principle. It is key to note that this set of planet attributes have never been found associated with another planet anywhere in our universe

- The earth is located in a solar system situated at a prime location within our Milky Way Galaxy that protects us from cosmic radiation. If our solar system was positioned in one of the spiral arms within our Milky Way, the earth would be continually pelted by levels of cosmic radiation harmful to life.

- The earth orbits our sun at a perfect distance to receive the right amount of constant light and energy from the sun. Additionally, our sun radiates the right color of light required for life on earth.

- The earth exists in a planetary solar system which includes giant planets that help shield the earth from meteors, comets, and other traveling space objects.

- The earth orbits the right kind of star necessary to support life; our sun radiates energy which is neither too strong nor too weak. Additionally, our sun is the right type of sun to allow photosynthesis on earth.

- The elliptical orbit of the earth around the sun and the rotational axis of the earth (23 degrees) enable the seasons and moderate temperatures needed to support life.

- The temperature of our oceans is at the right level for our water cycle to support life.

- The earth's moon stabilizes the tilt and rotation of our earth. Additionally, the moon orbits at the right distance from the earth in order to correctly impact ocean tides.

- The earth has a surrounding magnetic field which protects against excessive solar dust entering our atmosphere.

- The earth has an ozone layer to filter out much of the sun's harmful UV radiation.

- The earth maintains the right size and mass to support life. Even a very slight decrease or increase in the earth's size or mass would impact the force of gravity on the earth.

- Our earth retains an oxygen-rich atmosphere needed to support life. This atmosphere allows helpful radiation to hit the earth in the forms of sunlight and heat, but it stops excessive harmful radiation from hitting the earth.

- The earth's atmosphere preserves a balance of the right amounts of carbon dioxide, nitrogen, and oxygen to support plants, animals, and people. The atmosphere also possesses just the right density of elements to support life and provide livable temperatures at elevations where life exists.

- The earth's soil has the right amount of nutrients and minerals to support life.

- Our earth includes the right mixture of large bodies of water (oceans) and land (continents) that support a complex biosphere of life.

- The earth and its atmosphere sustains the right level of light reflectivity, which is the amount of light reflected off the earth versus the amount of light absorbed by the earth. A slightly greater reflectivity and the earth would become too cold; a slightly lesser reflectivity and the earth would become too hot.

Scientific Reason 17 Summary: It is not scientifically or logically reasonable to believe that random, designless, chance-based processes associated with evolution over billions of years are responsible for establishing the many precise and specifically calibrated earth attributes contributing to the Anthropic Principle.

⌘ Scientific Reason 18 ⌘ Oxidizing Atmosphere

The earth's atmosphere is known as an oxidizing atmosphere because it includes a significant amount (about 21%) of oxygen. This means our atmosphere will oxidize (break down) certain sensitive cell molecules (including DNA, proteins, and membranes) that are not protected inside a cell. Because of this oxidation process, some evolutionists have suggested that an early earth atmosphere billions of years ago was not an oxidizing atmosphere but rather a "reducing" atmosphere. A reducing atmosphere consists of very little or no oxygen. This idea of a reducing earth atmosphere has been proposed in order to accommodate the idea that the parts of the very first cell formed spontaneously in a primordial soup billions of years ago.

The concept of a reducing atmosphere becomes problematic in light of all geological data available today, because all data reveals that the earth's atmosphere has always been an oxidizing atmosphere comprised of significant amounts of oxygen. In fact, some research identifies that the earth's atmosphere thousands of years ago had an even greater percentage of oxygen than it has today. Consequently, the proposal of a reducing earth atmosphere cannot stand up to scientific observations. Geologists Erich Dimroth and Michael Kimberley report that they found no evidence to support a former earth atmosphere without oxygen:

◀ *In general, we find no evidence in the sedimentary distributions of carbon, sulfur, uranium or iron, that an oxygen-free atmosphere has existed at any time during the span of geological history recorded in well preserved sedimentary rocks.* ▶ [1]

Scientific Reason 18 Summary: All available geological data identify that the atmosphere of our earth has always contained significant amounts of oxygen, and thus, has never been able to support an oxygen-free environment required for the alleged spontaneous generation of a first cell.

⌘ Scientific Reason 19 ⌘ Atmospheric Helium

Because we scientifically observe atmospheric helium levels expected for an earth only thousands of years old, we now know that our atmosphere lacks the expected amount of helium for an earth billions of years old.

Physicist and creationist Melvin Cook, Ph.D., shared an important discovery in the journal *Nature*. He discovered that helium enters our atmosphere by two methods: blown in by solar winds and produced in the earth's crust by the radioactive decay of uranium. Based on the observed rates of helium build-up in our atmosphere, the present amount of helium would accumulate in less than 10,000 years. [2]

Atmospheric scientist and creationist Larry Vardiman, Ph.D., wrote an article addressing the reality that our atmosphere does not have enough helium as would be expected if the age of the earth was billions of years old, calculating that our atmosphere only has .04% of the helium expected after billions of years of accumulation. In the abstract to his article, he states:

◀ *If the earth is billions of years old, the radioactive production of helium in the earth's crust should have added a large quantity of helium to its atmosphere. Current diffusion models all indicate that helium escapes to space from the atmosphere at a rate much less than its production rate. The low concentration of helium actually measured would suggest that the earth's atmosphere must be quite young.* ▶ [3]

As an additional note, a team of creationists discovered that a huge amount of helium still exists within the earth's crust and mantle, with much of the helium still contained within crystalline minerals known as zircons. This proves significant because helium is an atom that can diffuse through solids very quickly, including the rock types within the earth's crust and mantle. If the earth's age was billions of years old, the helium would have already diffused out of the zircons, crust, and mantle. [4]

Scientific Reason 19 Summary: The amount of helium within our atmosphere and the amount of helium within our earth's crust and mantle match the amounts expected for an earth age of about 10,000 years. These scientific observations contradict the expected helium amounts that should be observed if evolution and its long age premise was true.

⌘ Scientific Reason 20 ⌘
Earth's Magnetic Field

Scientific research identifies the magnetic field of the earth as young in age, based on the fluctuations, rapid reversals, and observed decay of the earth's magnetic field strength.

Physicist and creationist David Russell Humphreys, Ph.D., explains some of the research that has been performed within the study of the earth's magnetic field strength. Referencing research that was published by Dr. Thomas Barnes in the 1970s, Dr. Humphreys found that the average intensity of the earth's magnetic field has decreased by about 7% since 1829 and that the total amount of energy in the earth's magnetic field has decreased by about 14% since 1829. [5]

The field of archaeomagnetism involves the study of the magnetic fields found in man-made objects. Archaeomagnetic data from locations all over the earth show that the intensity of the earth's magnetic field was about 40% greater in 1000 A.D. than it is today, and that it has steadily declined since about 1000 A.D. This amount of decay could not have been occurring for millions of years because, if it had, the earth's implied initial field strength millions of years ago would have been too massive to realistically exist and support life. [5]

Dr. Humphreys' research in paleomagnetism (the study of the amount of magnetism in rocks and geologic layers) has shown the earth's magnetic field strength has increased and decreased many times in the past. Dr. Humphrey's research has identified that, in spite of these increases and decreases, the field has been losing roughly half of its energy every 700 years. His research shows that the maximum (not actual) age of the earth is between 10,000 and 100,000 years. Regardless of the range of these two timeframes, this research certainly contradicts the age of the earth required by evolution, which proposes our earth to be 4.5 billion years old. [5]

Scientific Reason 20 Summary: The age of the earth's magnetic field, based on observation and scientific research, cannot be harmonized with evolution and its alleged billions of years of earth's history.

⌘ Scientific Reason 21 ⌘
Earth's Rotation and Orbit

Why does the earth consistently rotate? Just as significantly, why does the earth orbit our sun at a consistency and speed needed for life to exist on the earth?

Some scientists will say that the earth rotates because gases or explosions from the past caused it to spin, but this does not fully answer the question. Giving credit to past unknown gases or explosions does not explain where such gases or explosions came from in the first place, nor how these gases started perfectly rotating our earth, nor why the earth continues to rotate at the right speed needed to support life. Scientists do note that our earth's rotation rate is decreasing very slightly, but this has no impact that we know about today.

Some scientists will say the earth continues to spin due to the physics principle that an object with inertia will continue on its path until some outside force, or another object, has an effect upon it and stops it from moving. But we also know that perpetual motion does not exist, and all moving objects will eventually slow down and come to a stop. So the idea that inertia solves these questions is not true - it still does not explain how the earth continues to rotate with a consistency and speed required to support life.

Here are six interesting facts about the earth's rotation and its orbit around our sun:

- The earth rotates counter-clockwise once around its axis in 23 hours, 56 minutes and 4 seconds.

- A year on earth actually lasts 365.2464 days, which explains why we have a Leap Day on February 29th every 4 years.

- The earth is not a perfectly round sphere. It actually has a fairly consistent bulge around its equator caused by its spinning rotation. Consequently, the earth's circumference measures slightly wider at its equator than it does measuring from the north and south poles.

- The entire earth actually experiences periodic tiny wobbles, which slightly shift the earth on its axis. These wobbles are called nutation and occur mainly from the gravitational pulls of our sun and moon on the earth's equatorial bulge. These wobbles can continue for a long time, as one wobble can last as little as a few weeks or as long as several years.

- On the earth's surface its rotational speed measures about 1,000 miles per hour! Of course, we seldom stop to consider that we are living on an earth spinning this fast.

- The earth orbits the sun at about 67,000 miles per hour! We live on a speeding earth traveling around our sun, yet again we do not even notice this velocity.

Evolution holds that random and purposeless forces must be the sources that started and continually sustain both the consistent rotational speed as well as the orbital speed of our earth. These forces have supposedly been at work for billions of years. Such a belief is neither scientific nor observable.

Scientific Reason 21 Summary: The earth exhibits rotational consistency around its axis and orbital consistency around our sun. These attributes of the earth are specifically calibrated at the values needed to support life on earth. It is not scientifically, observationally, or logically reasonable to believe that evolution's random-chance and purposeless processes are the explanation for the earth's finely-tuned consistent rotation and orbit.

⌘ Scientific Reason 22 ⌘
Cycles

We see cycles all around us so often that we just take them for granted as normal and unimpressive. Many kinds of cycles exist - here are some significant examples:

- The water cycle that moves water from lakes and oceans to our atmosphere and then back to lakes and oceans

- The weather cycle within the earth's atmosphere

- The cycles of carbon and nitrogen through our biosphere

- The cycle of electrons around an atom's nucleus

- The cycle of minerals moving from soil to plants to animals and then eventually back to soil

Notably, cycles have no beginning or end. Additionally, without all aspects of a cycle existing simultaneously in perfect balance, the cycle will not be able to start, or if already in motion, to continue. Think of cycles as an all or nothing situation. A cycle cannot exist with only portions of it working, but rather all of its parts have to be fully available and fully working in order for the cycle to exist.

So cycles represent intention, design, purpose, and instantaneous success, all of which are just the opposite of what evolution could cause. Evolution stands on the belief that purposeless, unintelligent, designless, and random-chance processes are the cause of everything, and that these random chance-based processes supposedly occur through trial and error as slow, gradual, successive, and small physical changes over billions of years. Because all parts of a cycle must be fully available and fully working for the cycle to exist, it becomes impossible for evolution to be the source or cause of complex, interdependent cycles such as the water cycle, the weather cycle, the nitrogen cycle, the carbon cycle, the electron cycle, and the mineral cycle.

Biochemist and creationist George T. Javor, Ph.D., notes the significance of cycles:

◀ *Cycles do not have beginnings or ends. In order to bring them into existence, the forces responsible for the cycles have to be balanced, and if there are multiple steps required for the completion of a cycle, all of the components of the cycle have to be in place. Cycles speak of organization, of design, of rapid implementation, and of a designer.* ▶ [6]

Scientific Reason 22 Summary: It is not scientifically, observationally, or logically reasonable to propose that evolution, based on designless chance-based processes, could be responsible for the complex and interdependent cycles that exist on the earth.

⌘ Scientific Reason 23 ⌘
Rapid and Catastrophic Evidence

In many places around the world, the shape and appearance of land structures indicate formation by rapid and catastrophic liquid forces. Evolution rests upon the geological assumption of uniformitarianism, which in summary, proposes that all the geologic forces impacting the landscape of our earth have been occurring at basically the same rate and intensity that we see today.

As an example, when we look at the Colorado River in the Grand Canyon, the assumption of uniformitarianism would say that the river volumes seen today have been fairly consistent for the life of the river. But anyone who has ever viewed the Grand Canyon would reasonably ask if the massive signs of water destruction and erosion could really have occurred due to the water volumes observed in the Colorado River today. The sheer size of the Grand Canyon - stretching 277 miles long, spanning up to 18 miles wide, and dropping up to a mile deep - destroys any hypothesis implying the canyon was carved by a river the general size of the Colorado River.

Here are six catastrophic geological examples of water destruction and erosion that we can observe:

- There are many massive sedimentary layers that span hundreds of thousands of square miles in size and reach depths over a thousand feet. A powerful example is the Saint Peter Sandstone Megasequence, a layer of hardened sand in North America that is 3,000 by 1,000 miles wide and reaches several hundred feet deep. Catastrophic and complete flooding of the entire North American continent certainly explains such a layer, while the slow and gradual geological processes we observe today certainly do not.

- There are canyons around the world scoured hundreds of feet wide and deep, which are known as "over fit" canyons. They are over fit because they are massively larger than the size needed for the rivers flowing through them today. The rivers in them today do not demonstrate the water volume or erosive forces needed to create these canyons. Significant examples include the Copper Canyon in Mexico, the Tiger Leaping Gorge in China, and the Colca Canyon in Peru.

- There are scablands, such as those found in Washington, which signify a massive amount of water scoured the landscape in a very short period of time. The huge amounts of rushing water needed to blast away dirt and rock and create these scablands far exceeds the amount of water in these areas today.

- There are large, sharp and steep rock faces around the world, such as those at Yosemite National Park in California. Again, the amount of water, the uplifting forces, and the erosional forces needed to scour and shape these rocks no longer exist today.

- We see gigantic hardened lava flows such as the Columbia River Basalt Group spreading into sections of Washington, Oregon and Idaho. These hardened and stacked lava flows comprise an estimated 65,000 square miles with a volume of 40,000 cubic miles of lava. While small lava flows exist today, such as those in the Hawaiian Islands, we do not observe anything close to the gigantic lava flows observed in the Columbia River Basalt Group.

- Other reasons in this book highlight fossils, but note there are billions of fossils around the world that show signs they were buried rapidly and catastrophically within now-hardened sedimentary sand and rock layers. Obviously, all these living plants and animals did not have time to decay as plants and animals normally do, but rather, they were buried quickly and aggressively.

Geologist James H. Zumberge, Ph.D., provides a candid view about assuming that uniformitarianism is true:

◀ *From a purely scientific point of view, it is unwise to accept uniformitarianism as unalterable dogma... (One) should never close his mind to the possibility that conditions in past geological time were different than today.* ▶ [7]

Paleontologist and evolutionist Stephen Jay Gould, Ph.D., identifies that gradualism, another way to speak of uniformitarianism, is not observed in the geologic layers:

◀ *Gradualism, the idea that all change must be smooth, slow, and steady, was never read from the rocks. It was primarily a prejudice of nineteenth-century liberalism facing a world in revolution. But it continues to color our supposedly objective reading of life's history.* ▶ [8]

Geologist Edgar Heylmun, Ph.D., points out the reality that uniformitarianism has never been scientifically proven to exist:

◀ *It is hereby submitted that most scientists are guilty of an overly-zealous interpretation of the doctrine of uniformitarianism. Many instructors dismiss the possibilities of global catastrophes altogether, whereas others ridicule and scoff at the early ideas. These same instructors will implore their students to think scientifically and to develop the principles of multiple-working hypotheses. The fact is, the doctrine of uniformitarianism is no more 'proved' than some of the early ideas of world-wide cataclysms have been disproved.* ▶ [9]

Scientific Reason 23 Summary: The appearance of many geological landscapes on the earth display the results of rapid, aggressive, and catastrophic geologic forces that are not observable today. These landscapes fail to support slow and gradual forces that evolution and its billions of years require.

⌘ Scientific Reason 24 ⌘
Mississippi River Delta

On average, an estimated 600,000 cubic feet of water flows from the Mississippi River into the Gulf of Mexico each second of the day! Within this incredible amount of moving water, an estimated 300 million cubic yards of mud and sediment spill into the Gulf of Mexico each year. The current size of the Mississippi River delta extends 10,000 square miles and is increasing by 300 square miles each year. [10]

By observing the rate and mass of the annual mud and sediment accumulation in relation to the mass of the entire Mississippi River delta, the age of the delta calculates to about 4,000 years. Additionally, at the current growth rate of the delta, the entire Gulf of Mexico would have been filled with sediment in an estimated 10 million years. This, of course, does not mean that the Mississippi River delta is 10 million years old today, but it does tell us that the delta's age does not harmonize with evolution's billions of years. [11]

Scientific Reason 24 Summary: The scientifically observed current rate of mud and sediment accumulation into the Mississippi River Delta cannot be scientifically harmonized with evolution's requirement for billions of years.

⌘ Scientific Reason 25 ⌘
Ocean Sodium

Through scientific observation and measurements of the amount of sodium in our oceans, we can deduce the approximate age of our oceans.

Sodium enters our oceans by several methods:

- Sodium enters our oceans via volcanic dust flowing into our atmosphere and settling into the oceans.
- Sodium washes into our oceans by river erosion and glacier action.
- Sodium seeps into our oceans from the continental crust under the ocean.
- Sodium powers into the ocean through volcanic vents located at the bottom of the ocean.

Sodium primarily leaves the ocean via sea spray. Cresting waves create spray causing some of the wave water to evaporate quickly and leave trace amounts of sodium in the atmosphere.

Oceanographers have discovered that the amount of sodium entering our oceans is much greater than the amount of sodium leaving the oceans. Research shows that only about 27% of the sodium that enters the oceans ends up leaving the oceans, and the rest of the sodium remains in the oceans. Thus, our oceans are constantly growing "saltier". Oceanographers estimate that our oceans contain about 14.7 million billion tons of sodium. Each year, 457 million tons of sodium enter our oceans and 122 million tons of sodium leave our oceans. Thus, an estimated net gain of 335 million tons of sodium is added each year. [12]

Based on the known amount of sodium in our oceans and the known amount added each year, the maximum age of our oceans can be deduced. The maximum (not actual) age of our oceans - assuming they had no sodium originally and that there has never been any catastrophic events that added sodium - falls within tens of millions of years, which is a mere fraction of the evolutionary 4.5 billion-year age of the earth. [12]

Scientific Reason 25 Summary: The rate of sodium accumulation in our oceans provides observable scientific support that is incompatible with the billions of years of earth's history demanded by evolution.

⌘ Scientific Reason 26 ⌘
Ocean Sediment and Mud

Oceanographers now have accurate methods for measuring the ocean sediment growth in our oceans. By measuring the amount of sediment and mud eroding from our continents and settling into the bottoms of our oceans, this research provides another example of a scientific observation that contradicts evolution occurring over billions of years.

Due to rain erosion, melting snow runoff, frost erosion, wind erosion, and ocean waves pounding our shores, our oceans receive a continuous stream of sediments. Geologists estimate that 20 billion tons of sediment and mud erode off our continents and settle onto the ocean floor each year. Geologists estimate that a small portion of the sediment and mud cycles back into the earth's crust via the action of plate tectonics. An estimated 1-2 inches of sediment and mud cycles down into the earth's crust below the bottom of the ocean each year. This equals about one billion tons each year. So, net effect, the amount of sediment and mud at the bottom of our oceans is growing by about 19 billion tons each year! Additionally, scientists have measured the depths of sediment and mud settled in the bottom of our oceans. Some areas have very little sediment and mud while others have a depth of 2,000 feet, equating to an average depth of 1,200 feet. [13]

Based on the known amount of sediment and mud on the bottom of our oceans and the known amount added each year, the maximum age of our oceans can be deduced. The maximum age (not actual age) of our oceans - assuming there have never been any catastrophic events that significantly added sediment and mud - equates to 10 million years, which is a mere fraction of one percent of the alleged evolutionary age of the earth. The point here is not that the earth is 10 million years old, but rather that evolution promotes our earth has existed for 4.5 billion years but our ocean sediment and mud levels do not harmonize with these billions of years.

An interesting side note: at today's current rates of continental erosion, all the continents which exist today would erode down to sea level in only a few million years, which logically means the shapes and heights of our continents also do not harmonize with a billions-of-years earth age.

Scientific Reason 26 Summary: The scientifically observed rates of sediment and mud accumulation onto our ocean floors is not compatible with the billions of years of earth history demanded by evolution.

Chapter 3 Endnotes

1 – Erich Dimroth and Michael M. Kimberley, "Precambrian Atmospheric Oxygen: Evidence in the Sedimentary Distributions of Carbon, Sulfur, Uranium and Iron", *Canadian Journal of Earth Science*, Vol. 13 (1976) pages 1, 161-1, and 185.

2 – Melvin Cook, "Where is The Earth's Radiogenic Helium?", *Nature*, Vol. 179 (26 January 1957) page 213.

3 – Larry Vardiman, "The Age of the Earth's Atmosphere Estimated by its Helium Content", Presented at the First International Conference on Creationism, Pittsburgh, PA, August 4–9, 1986. Published by Creation Science Fellowship, Inc., Pittsburgh, PA, in the Proceedings of the First International Conference on Creationism, Editors R. E. Walsh, C. L. Brooks, and R. S. Crowell (1986) pages 187–194. <http://www.icr.org/article/age-earth-atmosphere-helium>

4 – David Russell Humphreys, Andrew A. Snelling, John Baumgardner, Steven A. Austin, "Helium Diffusion Rates Support Accelerated Nuclear Decay", *Answers in Depth*, Vol. 6 (2011). http://www.answersingenesis.org/articles/aid/v6/n1/accelerated-nuclear-decay>

5 – David Russell Humphreys, "The Earth's Magnetic Field is Young", *Acts & Facts*, Vol. 22 (1993) page 8. <http://www.icr.org/article/371>

6 – George T. Javor, *In Six Days – Why 50 Scientists Choose to Believe in Creation*, Editor John F. Ashton (Green Forest, AR: Master Books, 2001) page 135.

7 – James H. Zumberge, *Elements of Geology* (New York, Second Edition, John Wiley and Sons, 1963) page 200.

8 – Stephen Jay Gould, "An Early Start", *Natural History*, Vol. 87 (February 1978) page 24.

9 – Edgar B. Heylmun, "Should We Teach Uniformitarianism?", *Journal of Geological Education*, Vol. 19, No. 1 (January 1971) page 35.

10 – Douglas F. Kelly, *Creation and Change* (Scotland, UK: Christian Focus Publications, 1997) page 152.

11 – Benjamin Franklin Allen, "The Geologic Age of the Mississippi River", *Creation Research Society Quarterly Journal*, Vol. 9, No. 2 (September 1972). <http://www.creationresearch.org/crsq/abstracts/sum9_2.html>

12 – Steven A. Austin, David Russell Humphreys, "The Sea's Missing Salt: A Dilemma for Evolutionists", Presented at the Second International Conference on Creationism, Pittsburgh, PA, July 30–August 4, 1990. Published by Creation Science Fellowship, Inc., in the Proceedings of the Second International Conference on Creationism, Editors R. E. Walsh and C. L. Brooks (1990) pages 17–33. <http://www.icr.org/article/sea-missing-salt>

13 – Tas Walker, "The mud is missing", *Creation*, Vol. 32, Iss. 3 (July 2010) page 52. <http://creation.com/missing-mud>

Chapter 3 Resources

Anthropic Principle

- http://www.gotquestions.org/anthropic-principle.html
- http://www.compellingtruth.org/anthropic-principle.html
- http://thefactofcreation.blogspot.com/2012/08/anthropic-principle-the.html
- http://www.simpletoremember.com/articles/a/intelligent-design
- https://answersingenesis.org/astronomy/extrasolar-planets/just-right-for-life

Oxidizing Atmosphere

- http://www.icr.org/article/203
- http://cjes.geoscienceworld.org/content/13/9/1161.abstract
- http://www.darwinthenandnow.com/2013/10/oxygen-the-scourge-of-evolution-4
- https://answersingenesis.org/age-of-the-earth/the-first-atmosphere-geological-evidences-and-their-implications

Atmospheric Helium

- http://www.icr.org/article/age-earth-atmosphere-helium
- http://www.icr.org/article/young-helium-diffusion-age-zircons
- http://creation.com/blowing-old-earth-belief-away-helium
- http://creation.com/helium-in-the-earths-atmosphere
- http://creation.com/the-primitive-atmosphere
- http://www.answersingenesis.org/articles/aid/v6/n1/accelerated-nuclear-decay

Earth's Magnetic Field

- http://creation.com/the-earths-magnetic-field-evidence-that-the-earth-is-young
- http://www.answersingenesis.org/articles/cm/v13/n4/magnetic
- https://answersingenesis.org/astronomy/earth/the-earths-magnetic-field
- http://www.icr.org/article/depletion-earths-magnetic-field
- http://www.icr.org/article/371
- *Unformed and Unfilled* by Weston W. Fields

Rapid and Catastrophic Evidence

- http://www.icr.org/article/84/356

Mississippi River Delta

- http://www.creationresearch.org/crsq/abstracts/sum9_2.html
- *Creation and Change* by Christian Focus Publications

Ocean Sodium

- http://creation.com/salty-seas-evidence-for-a-young-earth
- http://www.icr.org/article/sea-missing-salt
- http://www.answersingenesis.org/assets/pdf/am/v7/n4/ECP-p879-881-sea-salt.pdf
- https://answersingenesis.org/evidence-for-creation/9-very-little-salt-in-the-sea

Ocean Sediment and Mud

- http://www.icr.org/research/index/researchp_lv_r01
- http://creation.com/missing-mud
- http://creation.com/eroding-ages
- https://answersingenesis.org/geology/sedimentation/1-very-little-sediment-on-the-seafloor

⌘ Chapter 4 ⌘

Fossils

Despite the bright promise that paleontology provides a means of 'seeing' evolution, it has presented some nasty difficulties for evolutionists, the most notorious of which is the presence of 'gaps' in the fossil record. Evolution requires intermediate forms between species and paleontology does not provide them. [2]

⌘ Scientific Reason 27 ⌘ Fossils in Sedimentary Layers

Billions of plants and animals fossils are found in sedimentary layers all over the world, with some fossils buried thousands of feet deep within a sedimentary layer. The existence of so many fossils cannot be due to the slow and gradual death of these plants and animals over very long periods of time. We all know that a dead creature will immediately begin to decay or be eaten by scavengers. So the fact that so many fossils exist all over the world, often buried in thousands of feet of hardened sand and mud layers, speaks to a rapid, catastrophic burial and fossilization processes occurring in massive proportions.

About 95% of all discovered fossils are marine creatures. Remarkably, marine fossils are found on every continent, even at the tops of mountains. This too speaks of a rapid, catastrophic, and massive event in which oceans once covered the entire earth. Remember that merely living in an ocean did not fossilize these creatures - the creatures that are living and dying in our oceans today of course are not being fossilized. Rather the discovered fossilized creatures were buried under massive amounts of sand, rock and mud while they were still alive in an ocean.

The observed fossil record does not match what evolution proposes. If evolution was true, which promotes a progression of advancing creatures that slowly, genetically, and physically morph into other creatures over millions of years, then the most "advanced" fossils would be in the top layers and the supposedly more "primitive" fossils would be in the bottom levels of the observed sedimentary layers. But in reality, geologists observe all types of fossils scattered throughout all sedimentary layers, and this fossil mixture actually occurs all over the earth. There are no places anywhere in the world where supposedly advanced fossils only exist within the top sedimentary layers and supposedly primitive fossils only exist in the bottom sedimentary layers!

Incredibly, scientific observation reveals gaps in the fossil record at all the places where transitional fossils between the major groups of creatures should be found. These major transitional fossil gaps occur between every distinction of creature types: between fish and amphibians, between amphibians and reptiles, between reptiles and birds, between reptiles and mammals, and between mammals and people. Paleontologist and evolutionist George Gaylord Simpson, Ph.D., notes this existence of gaps in the fossil record in his book *The Major Features of Evolution*:

◀ *It remains true, as every paleontologist knows, that most new species, genera and families, and that nearly all categories above the level of families, appear in the [fossil] record suddenly and are not led up to by known, gradual, completely continuous, transitional sequences.* ▶ [1]

In the journal *Evolution*, paleontologist and evolutionist David B. Kitts, Ph.D., calls out the existence of these fossil gaps:

◀ *Despite the bright promise that paleontology provides a means of 'seeing' evolution, it has presented some nasty difficulties for evolutionists, the most notorious of which is the presence of 'gaps' in the fossil record. Evolution requires intermediate forms between species and paleontology does not provide them.* ▶ [2]

Paleontologist and evolutionist David M. Raup, Ph.D., notes the lack of fossil evidence for progressing creatures in the journal *Science*, within an article he titled *Evolution and the Fossil Record*:

◀ *In the years after Darwin, his advocates hoped to find predictable progressions. In general, these have not been found - yet the optimism has died hard, and some pure fantasy has crept into textbooks.* ▶ [3]

In his book *The Problems of Evolution*, zoologist and evolutionist Mark Ridley, Ph.D., highlights the lack of fossil lineages in sedimentary layers. He also notes that Charles Darwin himself could not identify even one such example of a transitional fossil:

◀ *The fossil record of evolutionary change within single evolutionary lineages is very poor. If evolution is true, species originate through changes of ancestral species: one might expect to be able to see this in the fossil record. In fact it can rarely be seen. In 1859 Darwin could not cite a single example.* ▶ [4]

Paleontologist and evolutionist Derek Ager, Ph.D., notes that geological observations of the fossil record never display gradual changes in creatures; rather, we simply observe the appearance and disappearance of entire groups of creatures:

◀ *The point emerges that, if we examine the fossil record in detail, whether at the level of orders or of species, we find - over and over again - not gradual evolution, but the sudden explosion of one group at the expense of another.* ▶ [5]

Charles Darwin himself actually notes the lack of intermediate forms within the fossil record:

◀ *The number of intermediate varieties which have formerly existed on earth must be truly enormous. Why then is not every geological formation and every stratum full of such intermediate links? Geology assuredly does not reveal any such finely graduated organic chain; and this, perhaps, is the most obvious and gravest objection which can be urged against my theory.* ▶ [6]

Paleontologist and evolutionist Stephen Jay Gould, Ph.D., also notes the lack of intermediate fossils in the fossil record:

◀ *The absence of fossil evidence for intermediary stages between major transitions in organic design, indeed our inability, even in our imagination, to construct functional intermediates in many cases, has been a persistent and nagging problem for gradualistic accounts of evolution.* ▶ [7]

Chemist and creationist Duane Gish, Ph.D., comments on the lack of transitional fossils:

◀ *This discontinuity is so complete and so extensive that this evidence alone is sufficient to disprove evolution... Evolution required continuity and here it fails absolutely. Let us return to where there are no disputes, no questions, no doubts about the actual evidence - the total absence of ancestors and transitional forms for the complex invertebrates and for each class of fish. The gaps are so immense and so complete that there is no doubt that these creatures all appeared fully formed, with no ancestors. These facts completely demolish Darwinian evolution.* ▶ [8]

Geneticist and evolutionist Nils Heribert-Nilsson, Ph.D., performed a 40-plus year genetics experiment to try to discover support for evolution, but at the end of his experiment he finally concludes that the fossil record does not support evolution:

◀ *My attempts to demonstrate evolution by an experiment carried on for more than 40 years have completely failed... it may be firmly maintained that it is not even possible to make a caricature of an evolution out of paleo-biological facts. The fossil material is now so complete... and the lack of transitional series cannot be explained as being due to the scarcity of material. The deficiencies are real, they will never be filled.* ▶ [9]

Scientific Reason 27 Summary: Scientific research demonstrates three truths:

1) A mixture of supposedly "advanced" and "primitive" fossilized creatures exist throughout all sedimentary layers all over the earth.
2) There is a complete lack of any conclusive and observable fossilized transitional creatures.
3) There are gaps in the fossil record between all major groups of creatures.

Because all three of these significant scientific observations are true, evolution and its idea that the fossil record displays evolving creatures is scientifically unsubstantiated.

⌘ Scientific Reason 28 ⌘
Number of Transitional Human Fossils

If evolution really occurred over a past 2.5 million years, we should find billions of transitional human and pre-human fossils. There are no conclusive examples. Rather, evolutionary scientists promote a small number (less than 100) of alleged transitional human and pre-human fossils, and all of these are hotly debated as to their authenticity.

An amazing scenario plays out here. Proponents of evolution will promote various fossils as transitional forms of human evolution, and the opponents of evolution will identify those same fossils as frauds or other creatures. Regardless of the side anyone takes on the credibility of these fossils, there are less than 100 supposed transitional fossils that are hotly debated. But here is the key to this reason: if human evolution had occurred for the last 2.5 million years, scientifically and mathematically there would be trillions upon trillions upon trillions of transitional forms of humans and pre-humans (See Scientific Reason 64), not merely a handful of hotly debated fossils.

Another key point here: if human evolution was true, there would be transitional forms of people living today, but there are none. Sadly, over the last few hundred years some people groups, such as the Aborigines of Australia, have been wrongly labeled as examples of transitional humans, but this thinking has been shown to be completely false. Photo-journalist John Reader distinctively points out the lack of human transitional forms:

◀ *The entire hominid collection known today would barely cover a billiard table... the collection is so tantalizingly incomplete, and the specimens themselves often so fragmented and inconclusive, that more can be said about what is missing than about what is present... but ever since Darwin's work inspired the notion that fossils linking modern man and extinct ancestor would provide the most convincing proof of human evolution, preconceptions have led evidence by the nose in the study of fossil man.* ▶ [10]

Anthropologist and evolutionist Lyall Watson, Ph.D., also highlights this lack of human transitional forms in a *Science Digest* article:

◀ *The fossils that decorate our family tree are so scarce that there are still more scientists than specimens. The remarkable fact is that all the physical evidence we have for human evolution can still be placed, with room to spare, inside a single coffin... Modern apes, for instance, seem to have sprung out of nowhere. They have no yesterday, no fossil record. And the true origin of modern human, of upright naked, toolmaking, big-brained beings, is if we are to be honest with ourselves, an equally mysterious matter.* ▶ [11]

Anthropologist and evolutionist Tim White, Ph.D., identifies the motivation for the announcements of so many fraudulent human transitional fossils:

◀ *The problem with a lot of anthropologists is that they want so much to find a hominid that any scrap of bone becomes a hominid bone.* ▶ [12]

Genetics professor J. S. Jones, in his review of the book *A Thousand and One Eves*, highlights how paleontologists have recently chose to ignore this lack of transitional fossils:

◀ *Paleoanthropologists seem to make up for a lack of fossils with an excess of fury, and this must now be the only science in which it is still possible to become famous just by having an opinion. As one cynic says, in human paleontology the consensus depends on who shouts loudest.* ▶ [13]

British Museum of Natural History Director Collin Patterson, Ph.D., wrote a book titled *Evolution*. Creationist Luther Sunderland wrote Dr. Patterson, asking why he did not include any photographs of a transitional fossil in his book. In response, Dr. Patterson wrote back and said the following to Mr. Sunderland, which Mr. Sunderland documented in his book *Darwin's Enigma*:

◀ *I fully agree with your comments on the lack of direct illustration of evolutionary transitions in my book. If I knew of any, fossil or living, I would certainly have included them. You suggest that an artist should be used to visualise such transformations, but where would he get the information from? I could not, honestly, provide it, and if I were to leave it to artistic licence, would that not mislead the reader? Yet Gould [Stephen J. Gould] and the American Museum people are hard to contradict when they say there are no transitional fossils... You say that I should at least 'show a photo of the fossil from which each type of organism was derived.' I will lay it on the line - there is not one such fossil for which one could make a watertight argument.* ▶ [14]

Scientific Reason 28 Summary: If evolution was true, trillions upon trillions upon trillions (See Scientific Reason 64) of transitional human and pre-human fossils would exist. The fact that they do not exist tells us that evolution over long ages is scientifically unsubstantiated and not true.

⌘ Scientific Reason 29 ⌘
Human Artifacts in the Fossil Record

Evolution includes two foundational beliefs. First, the earth's age extends back 4.5 billion years. Second, people began to evolve from apes into supposed early hominids (part ape and part human) 2.5 million years ago.

Contrary to the timeframes promoted by evolutionary beliefs, human artifacts have been found in coal layers and sedimentary layers dated to be hundreds of millions of years old. One artifact example is a metal bell that was found in a West Virginia coal layer dated at 300 million years old. If evolving forms of people have only been on earth for the last 2.5 million years, then, of course, this man-made object could not exist some 300 million years ago. A link in the resources for this reason provides more details about this amazing anomalous artifact. [15]

Many other examples of man-made objects located within coal and hardened sandstone layers, supposedly hundreds of millions of years old, have been discovered over the last few hundred years. The list includes: boats, jewelry, knives, hammers, pots, and other tools. See the link in the chapter resources for this reason for more examples of artifact discoveries that contradict the evolutionary timeframe.

So, an obvious question surfaces here: if people have been around "only" 2.5 million years, then how could these artifacts exist in coal and hardened sandstone layers dated to be tens and hundreds of millions of years old? A clear answer emerges: the layers are not really tens or hundreds of millions of years old.

Scientific Reason 29 Summary: The observable existence of human artifacts found within layers of coal and hardened sandstone reveals that the layers are not really tens or hundreds of millions of years old as proposed by evolution and its dependence upon long ages.

⌘ Scientific Reason 30 ⌘
Cambrian Explosion

Within the fossil record, the phrase "Cambrian explosion" describes the sudden appearance of fully-developed fossilized creatures which do not have any transitional fossils in the layers below or above them. The Cambrian explosion displays a large set of purportedly "advanced" fossilized creatures and includes all phylas of creatures known to exist. This is just the opposite of what should exist if evolution was true. Evolution requires a continuum or progression from primitive to advanced fossils starting with the lowest sedimentary layer and progressing to the top sedimentary layer, but the fossils in the Cambrian explosion display just the opposite. The fossils in the Cambrian explosion suddenly appear as fully formed creatures, and many of these creatures are actually still alive in the same shape and form today! Geologists observe that the fossils in the layers immediately leading up to the Cambrian explosion do not display any evolutionary changes. And just as significant, in the sedimentary layers above the Cambrian explosion, the fossilized creatures there also do not display any progressing evolutionary structures.

Paleontologist and evolutionist Stephen Jay Gould, Ph.D., highlights that the creatures observed in the Cambrian explosion do not have any transitional creatures that lead up to them, not even one:

Studies that began in the early 1950s and continue at an accelerating pace today have revealed an extensive Precambrian fossil record, but the problem of the Cambrian explosion has not receded, since our more extensive labor has still failed to identify any creature that might serve as a plausible immediate ancestor for the Cambrian faunas. [16]

Atheist and evolutionist Richard Dawkins also notes that the creatures observed in the Cambrian explosion do not display any forms of evolutionary ancestors:

The Cambrian strata of rocks, vintage about 600 million years [evolutionists are now dating the beginning of the Cambrian at about 530 million years], are the oldest in which we find most of the major invertebrate groups. And we find many of them already in an advanced state of evolution, the very first time they appear. It is as though they were just planted there, without any evolutionary history. [17]

Charles Darwin knew about the Cambrian explosion and admitted it cannot synchronize with his suggestion of evolution. Additionally, he recognizes that the Cambrian explosion actually provides support that his suggestion of evolution is invalid:

There is another and allied difficulty, which is much more serious. I allude to the manner in which species belonging to several of the main divisions of the animal kingdom suddenly appear in the lowest known fossiliferous rocks... The abrupt manner in which whole groups of species suddenly appear in certain formations, has been urged by several paleontologists - for instance, by Agassiz, Pictet, and Sedgwick - as a fatal objection to the belief in the transmutation of species. If numerous species, belonging to the same genera or families, have really started into life at once, the fact would be fatal to the theory of evolution through natural selection... To the question why we do not find rich fossiliferous deposits belonging to these assumed earliest periods prior to the Cambrian system, I can give no satisfactory answer... The case at present must remain inexplicable, and may be truly urged as a valid argument against the views here entertained. [18]

Scientific Reason 30 Summary: The observable existence of the Cambrian explosion within sedimentary layers, in which creatures appear as fully formed fossilized creatures with no accompanying transitional fossils leading up to them and no advanced fossils appearing in the layers above them, displays exactly the opposite of what should be observed according to evolution and its premise of the slow and gradual development of new species through transitional forms.

⌘ Scientific Reason 31 ⌘ Direct Dinosaur Ancestor Fossil

About 30 million dinosaur bones have been discovered to date, and paleontologists are currently familiar with about 700 dinosaur species. Yet not one fossil of a direct ancestor has been found for any dinosaur, which some evolutionists suggest appeared as a small, crocodile-like animal. Also, the supposed "common ancestor" of all dinosaurs has never been found. A scientific reality is that all discovered dinosaur fossils appear fully formed and not transitional. [19]

Medical doctor and creationist Jerry Bergman, M.D., sums up the scientific reality of the lack of any dinosaur ancestor fossils:

◀ *The evidence for dinosaur evolution was reviewed, along with the various theories of dinosaur evolution and the evidence for their support. Dinosaurs are commonly believed to have evolved from a small, crocodile-like animal; however, a review of the known fossils provides no evidence for dinosaur evolution from non-dinosaurs, despite the excellent and abundant dinosaur record.*

Over 30 million dinosaur bones and parts, some in excellent states of preservation, have been identified, and although much speculation exists, not a single documented plausible direct ancestor has yet been located. All known dinosaurs appear fully formed in the fossil record. ▶ [20]

Scientific Reason 31 Summary: The complete lack of any conclusive scientific evidence of a fossil ancestor for any of the 30 million dinosaur bones discovered to date, which represent about 700 dinosaur species, is not compatible with evolution and its premise that dinosaurs evolved from less advanced creatures living millions of years ago.

⌘ Scientific Reason 32 ⌘
"Living" Fossils

A "living" fossil is an evolutionary expression used to identify fossilized creatures that allegedly lived millions of years ago but in reality are living today in the same shape and size.

Some examples of living fossils include: the coelacanth fish, the horseshoe crab, the cockroach, the crocodile, the nautilus, the lungfish, the sturgeon, the mackerel, the freshwater bass, the herring, the needlefish, the lobster, the crawfish, the shark, the jellyfish, the sea sponge, the frog, the bee, the ant, the butterfly, the termite, the dragonfly, the salamander, the spider, the millipede, the gingko tree, and the Wollemi Pine tree. It is important to note that these supposed living fossils have been found within many of the sedimentary layers that are given dates going back hundreds of millions of years. See the resources for this reason for an easy-to-read chart that highlights many of the creatures and the sedimentary layers in which they have been found.

Biologist and creationist Carl Werner, Ph.D., has done extensive research into living fossils and wrote a book called *Living Fossils – Evolution: the Grand Experiment*. He has traveled for years and taken thousands of pictures of these living fossils. He found fossils of animals that are alive today and were buried in the very same sedimentary layers as dinosaur fossils. Note that if evolution was true, fossils of dinosaurs and animals (still living today) would never be found in the same sedimentary layer because, according to evolutionary beliefs, dinosaurs lived tens of millions of years before modern animals. In an interview, Dr. Werner highlights:

◀ *We found fossilized examples from every major invertebrate animal phylum living today including: arthropods (insects, crustaceans etc.), shellfish, echinoderms (starfish, crinoids, brittle stars, etc.), corals, sponges, and segmented worms (earthworms, marine worms). The vertebrates - animals with backbones such as fish, amphibians, reptiles, birds and mammals - show this same pattern.* ▶ [21]

Dr. Carl Werner also highlights that even though almost 100 complete mammal skeletons have been discovered in the same layers where dinosaur skeletons have been found, most museums will seldom display these complete mammal skeletons due to their unscientific bias towards promoting evolution. He states:

◀ *Few are aware of the great number of mammal species found with dinosaurs. Paleontologists have found 432 mammal species in the dinosaur layers; almost as many as the number of dinosaur species. These include nearly 100 complete mammal skeletons. But where are these fossils? We visited 60 museums but did not see a single complete mammal skeleton from the dinosaur layers displayed at any of these museums. This is amazing. Also, we saw only a few dozen incomplete skeletons/single bones of the 432 mammal species found so far. Why don't the museums display these mammal fossils and also the bird fossils?* ▶ [21]

In consideration of Dr. Werner's research and discoveries, a question becomes relevant: if evolution was true, why would the creatures living today be the very same body shape and size that supposedly existed tens or hundreds of millions of years ago? Of course, the scientific and logical answer to this is that all these creatures are not evolving over tens or hundreds of millions of years as evolution proposes.

Scientific Reason 32 Summary. The observable existence of hundreds of living fossils is not compatible with evolution and its premise that creatures living today evolved from less advanced creatures which lived hundreds of millions of years ago.

⌘ Scientific Reason 33 ⌘
Polystrate Fossils

In many sedimentary layers of hardened sandstone, geologists observe polystrate fossils. A polystrate fossil exists in two or more layers of rock strata, and a common example is a fossilized tree trunk positioned perpendicularly within several horizontal layers. Because we can observe polystrate fossils existing in two or more layers, we can deduce that both the polystrate fossils and the layers themselves cannot be millions of years old.

Here is why the multiple rock strata layers and the fossilized vertical object cannot be millions of years old. During the time when only the first layer existed and the tree existed partially within it, the portion of the tree sticking out of the layer would have been exposed to weather, insects, and microorganisms. Thus, the protruding portion would have decayed in only a few years. The fact that the tree exists undecayed and fossilized in two or more layers means the layers were formed very quickly in relation to each another, even possibly at the same time. Thus, polystrate fossils only make sense scientifically if they were completely buried and fossilized very quickly, before any rot, insect impact, or decay occurred.

In his book *The New Catastrophism*, paleontologist and evolutionist Derek Ager, Ph.D., offers scientific insights why polystrate tree fossils found in the British Coal Measures do not support millions of years required by evolution:

◀ *If one estimates the total thickness of the British Coal Measures as about 1,000 meters, laid down in about 10 million years, then, assuming a constant rate of sedimentation, it would have taken 100,000 years to bury a tree 10 meters high, which is ridiculous.* ▶ [22]

Scientific Reason 33 Summary: The existence of polystrate fossils in the fossil record is not scientifically compatible with evolution and its reliance upon time spans of millions of years.

⌘ Scientific Reason 34 ⌘ Petrification

Evolution proposes that billions of years have passed. To support this idea, proponents of evolution often highlight that objects must take millions of years to petrify. Actually, today we have numerous examples of quick petrification: objects such as wood, hats, boots, and teddy bears hardening and petrifying into rock in less than 100 years. Resources of quick-petrification examples are included at the end of this chapter.

Millions of years are not needed for objects to become petrified. Rather, petrification can occur under the right conditions in only a few years or decades. The petrification process occurs when an object becomes immersed within water containing silica or similar minerals, enabling the silica or minerals to fill the microscopic pores and cavities of the object, which in turn causes the object's soft tissues to dissolve and become replaced with the minerals. Due to the recent scientific discovery of this process, we now know that objects can petrify in only a few years or decades.

Scientific Reason 34 Summary: We now have scientific evidence that all kinds of objects, whether composed of wood, cloth, or once-living tissue, can petrify in as little as a few years or decades. Therefore, evolution's proposal that petrification supports millions of years is wrong.

⌘ Scientific Reason 35 ⌘ Soft Tissues in Fossils

In the last 50 years, a large number of fossils which contain soft tissues have been discovered. The list of soft tissue types is significant and includes: DNA, hemoglobin, skin, keratin, cartilage, collagen, bone marrow, melanocytes, blood vessels, red blood cells, muscle, hair, retina, feathers, and proteins.

This significant observable scientific evidence means that these fossils are not millions of years old, because logically and scientifically none of these soft tissue types could survive in an unfossilized soft state for millions of years. In fact, to emphasize the point, these soft tissues could remain soft for only thousands of years at most, because of the way tissues break down at cellular and molecular levels.

Most soft tissues will chemically break down within a matter of days. For a soft tissue to last even a thousand years is extremely rare. Soft tissues will take longer than a few days to break down if they are buried in a substance that greatly reduces their decomposition rate, such as hardened rock or ice. But even if soft tissues are buried in rock or frozen in ice, the tissues are not immune to chemical and molecular breakdown, and consequently, it is chemically impossible for soft tissues to last millions of years.

If evolution was true, none of the many soft tissue discoveries within fossils would be occurring. The fact that many soft tissue discoveries are occurring tells us that these creatures had to live, at most, only a few thousand years ago. On its website, ICR has compiled a significant list of soft tissue discoveries from the last 50 years. The list includes the year of discovery and the publication in which the discovery was announced. It is a fascinating list which can be found in the chapter resources for this reason.

One of the most publicized and significant examples of soft tissue discovery belongs to paleontologist Mary Schweitzer, Ph.D., and her scientific team. They discovered red blood cells and blood vessels within a Tyrannosaurus Rex bone which was supposedly 68 million years old. During an interview, Dr. Schweitzer highlights the significance of her discovery:

◀ *If you take a blood sample, and you stick it on a shelf, you have nothing recognizable in about a week. So why would there be anything left in dinosaurs?* ▶ [23]

Scientific Reason 35 Summary: The observable discoveries of soft tissues within fossils, which at chemical and molecular levels cannot be more than a few thousand years old, scientifically identifies that these fossils are not tens or hundreds of millions of years old as proposed by evolution and its assumption of long ages.

Chapter 4 Endnotes

1 – George Gaylord Simpson, *The Major Features of Evolution* (New York: Columbia University Press, 1953) page 360.

2 – David B. Kitts, “Paleontology and Evolutionary Theory”, *Evolution*, Vol. 28 (September 1974) pages 458-472.

3 – David M. Raup, “Evolution and the Fossil Record”, *Science*, Vol. 213 (17July1981) page 289.

4 – Mark Ridley, *The Problems of Evolution* (New York: Oxford University Press, 1985) pages 11 and 159.

5 – Derek Ager, “The Nature of the Fossil Record”, *Proceedings of the Geological Association*, Vol. 87, No. 2 (1976) page 133. Presidential address delivered on 5March1976.

6 – Charles Darwin, *On the Origin of Species by Means of Natural Selection, or the Preservation of Favoured Races in the Struggle for Life* (1859) page 323.

7 – Stephen Jay Gould, "Is a new and general theory of evolution emerging?”, *Paleobiology*, Vol. 1 (January 1980) page 127.

8 – Duane Gish, *Letter to a Theistic Evolutionist* (Southfield, MI: ICON Publishing Group, 2012) pages 51 and 62.

9 – Nils Heribert-Nilsson, *Synthetic Speciation* (Sweden: Verlag CWK Gleerup, 1953) page 31.

10 – John Reader, "Whatever happened to Zinjanthropus?”, *New Scientist* (26March1981) page 802.

11 – Lyall Watson, “The Water People”, *Science Digest*, Vol. 90 (May 1982) page 44.

12 – Ian Anderson, "Hominoid collarbone exposed as dolphin's rib", *New Scientist* (28April1983) page 199.

13 – J. S. Jones, “A Thousand and One Eves, a review of *The Search for Eve”, Nature*, Vol. 345 (31May1990) pages 395-396.

14 – Luther Sunderland, *Darwin’s Enigma* (Green Forest, AR: Master Books, 1998) pages 101–102. Patterson’s letter was written in 1979.

15 – John D. Morris, “An Amazing Anomalous Fossil”, *Acts & Facts*, Vol. 39, Iss. 2 (2010) page 16. <http://www.icr.org/article/amazing-anomalous-fossil>

16 – Stephen Jay Gould, “A Short Way to Big Ends”, *Natural History*, Vol. 95 (January 1986) page 18.

17 – Richard Dawkins, *The Blind Watchmaker* (New York: W. W. Norton, 1987) page 229.

18 – Charles Darwin, *On the Origin of Species by Means of Natural Selection, or the Preservation of Favoured Races in the Struggle for Life* (1859) pages 344-351.

19 – Carl Werner, *Evolution the Grand Experiment* (Green Forest, AR: New Leaf Press, 2007) page 128.

20 – Jerry Bergman, “The Evolution of Dinosaurs: Much Conjecture, Little Evidence”, Creationresearch.org, Web 18September2015. <http://www.creationresearch.org/crsq/articles/46/46_2/CRSQ%20Fall%2009%20Bergman.pdf>

21 – Don Batten, “Living fossils: a powerful argument for creation”, *Creation*, Vol. 33, Iss. 2 (April 2011). <http://creation.com/werner-living-fossils>

22 – Derek Ager, *The New Catastrophism* (Cambridge, England: Cambridge University Press, 1993) page 49.

23 – Barry Yeoman, “Schweitzer's Dangerous Discovery”, Discovermagazine.com, Pub 27April2006. Web 18September2015. <http://discovermagazine.com/2006/apr/dinosaur-dna>

Chapter 4 Resources

Fossils in Sedimentary Layers

- https://answersingenesis.org/fossils/how-are-fossils-formed/how-fast
- http://www.icr.org/fossils-stasis
- http://creation.com/fossils-questions-and-answers
- *In The Beginning* at http://www.creationscience.com/onlinebook
- *The New Answers Book 2: Chapter 31* by Answers in Genesis
- *Evolution's Achilles Heel: Chapter 4* by Creation Ministries International
- *Evolution the Grand Experiment: Chapter 6* by Carl Werner
- *Untold Secrets of Planet Earth – Flood Fossils* by Vance Nelson

Number of Transitional Human Fossils

- http://creation.com/abandoned-transitional-forms
- http://creation.com/fossils-questions-and-answers#missing
- http://www.icr.org/article/1103/268
- *The Creation Answers Book: Chapter 15* by Creation Ministries International

Human Artifacts in Fossils Record

- http://www.icr.org/article/amazing-anomalous-fossil (Anomalous Artifact)
- http://www.icr.org/article/possible-human-artifact-found-coal (Artifact Discoveries)
- http://www.icr.org/article/fossil-discoveries-disrupt-evolutionary
- http://www.genesispark.com/exhibits/evidence/paleontological/artifacts/hammer
- http://www.hissheep.org/evolution/the_geologic_column_exposed_and_debunked.html

Cambrian Explosion

- http://creation.com/exploding-evolution
- http://creation.com/cambrian-explosion-and-animal-biodiversity
- https://answersingenesis.org/fossils/fossil-record/one-lifes-unexpected-explosion
- http://www.icr.org/article/cambrian-explosion-problem-solved

Direct Dinosaur Ancestor Fossils

- http://www.creationresearch.org/crsq/articles/46/46_2/CRSQ%20Fall%2009%20Bergman.pdf
- http://creation.com/origin-demise-dinosaurs
- http://creation.com/dinosaur-fairy-tales
- https://answersingenesis.org/dinosaurs
- *The Great Dinosaur Mystery Solved!* by Master Books
- *Evolution the Grand Experiment: Chapter 12* by Carl Werner

"Living" Fossils

- https://cdn-assets.answersingenesis.org/doc/articles/am/v6/n1/fossil-chart.pdf (Living Fossils Chart)
- http://www.icr.org/living-fossils
- http://creation.com/living-fossils-enigma
- http://creation.com/werner-living-fossils
- https://answersingenesis.org/fossils/living-fossils/living-fossils-coelecanth-darwin
- *Creation Answer Book* by Creation Ministries International

Polystrate Fossils

- http://www.icr.org/article/1144/274
- http://www.icr.org/article/classic-polystrate-fossil
- http://creation.com/polystrate-fossils-evidence-for-a-young-earth
- http://creation.com/early-arguments-for-deep-time-1
- https://answersingenesis.org/fossils/fossil-record/the-fossil-record-1

Petrification

- http://www.icr.org/article/569/274 (Petrification Examples)
- *Evolution's Achilles Heel* by Creation Ministries International
- https://answersingenesis.org/fossils/how-are-fossils-formed/instant-petrified-wood

Soft Tissues in Fossils

- http://www.icr.org/soft-tissue-list (Soft Tissue Discoveries)
- http://creation.com/schweitzers-dangerous-discovery
- http://discovermagazine.com/2006/apr/dinosaur-dna
- https://answersingenesis.org/dinosaurs/bones/more-soft-tissue-in-old-fossils
- https://answersingenesis.org/dinosaurs/bones/two-those-not-so-dry-bones
- *In Six Days – Why 50 Scientists Choose to Believe in Creation* by Master Books
- *Creation Basics and Beyond: Chapter 40* by Institute For Creation Research

⌘ Chapter 5 ⌘

Geology

The intelligent layman has long suspected circular reasoning in the use of rocks to date fossils and fossils to date rocks. The geologist has never bothered to think of a good reply, feeling the explanations are not worth the trouble as long as the work brings results. This is supposed to be hard-headed pragmatism. [4]

⌘ Scientific Reason 36 ⌘
Geologic Column

A chart of a geologic column is commonly offered as support for evolution. Nearly everyone has seen it, as it shows up in school textbooks, museums, and scientific articles. It usually appears as a diagram showing 13 geologic layers, with the layers given "historic" names such as Paleozoic, Cenozoic, and Mesozoic. Each layer may be labeled with the fossilized creatures that supposedly have been found only in that specific layer, and the creatures displayed usually progress from the most "primitive" ones displayed in the bottom layers to the more "advanced" ones displayed in the top layers of the chart. There are three significant scientific problems with the proposed geologic column that add up to the scientific reality that it does not physically exist anywhere in the world!

The first problem is that in hundreds of places around the world, the layers of the geologic column occur in the wrong order. If each layer really represented a timeframe in the earth's historical past, then logically, the layers would always appear in the same sequence. Consequently, these "disordered" geologic layers destroy the validity of the geologic column. See the Geologic Column resources in the chapter resources for this reason for more in-depth info on this and the next two problems.

A second serious problem is that layers are missing in 99.6% of the locations around the world where the column is claimed to appear. Geologist and creationist John Woodmorappe, Ph.D., reports in an ICR article:

◀ *Approximately 77% of the earth's surface area on land and under the sea has seven or more (70% or more) of the strata systems missing beneath; 94% of the earth's surface has three or more systems missing beneath; and an estimated 99.6% has at least one missing system.* ▶ [1]

A third scientific problem is that the creatures drawn within specific chart layers are often found, in reality, within other geologic layers. Geologists actually observe that all types of fossils are dispersed throughout multiple layers. There are no bottom layers with only primitive creatures and no upper layers with only advanced creatures. This lack of "upward evolutionary progression" also undercuts the scientific validity of the geologic column.

Scientific Reason 36 Summary: The geologic column fails to scientifically support evolution and its required billions of years of earth history three ways: 1) Observable geological evidence reveals that real-life geologic layers exist in different sequences compared to geologic chart drawings. 2) Geologic layers are missing in 99.6% of the locations where the geologic column is proposed to exist. 3) Fossilized creatures are actually found in other geologic layers contrary to geologic chart drawings.

⌘ Scientific Reason 37 ⌘
Dating of Geologic Layers

We observe horizontal layers of sedimentary rock in many places and on every continent. Naturally, we all wonder how old the layers are and when they formed. Unfortunately, the dating of sedimentary rock layers and the fossils within them is often performed using non-credible scientific methods that involve circular reasoning.

Real science involves observing objects or situations, forming a hypothesis, and researching and performing tests in order to validate or reject the hypothesis. Sadly, the proponents of evolution sometimes do not follow this established scientific method to determine the ages of sedimentary layers or the fossils found in them. Rather, when an age for a geologic layer becomes accepted (by their peers), some scientists interpret the age of fossils found within that layer to be the same age. Just as unscientifically, an age assigned to a fossil leads some scientists to assume the age of the geologic layer is the same as the fossil age.

The point here is that this method of thinking - dating geologic layers and the fossils within them according to personal bias or expectation - certainly fails to involve actual scientific observation and testing that the scientific method requires. In addition to the issue of bias, this reasoning method is circular: the layers are used to date the fossils and the fossils are used to date the layers. The unscientific use of circular reasoning has been known for years and was even highlighted in the *Encyclopedia Britannica* back in the 1940s:

◀ *It cannot be denied that from a strictly philosophical standpoint geologists are here arguing in a circle. The succession of organisms has been determined by a study of their remains embedded in the rocks, and the relative ages of the rocks are determined by the remains of organisms that they contain.* ▶ [2]

Paleontologist and evolutionist Ronald West, Ph.D., offers a candid assessment why this type of unscientific thinking should not be used to identify the age of geologic layers:

◀ *Contrary to what most scientists write, the fossil record does not support the Darwinian theory of evolution because it is this theory (there are several) which we use to interpret the fossil record. By doing so we are guilty of circular reasoning if we then say the fossil record supports this theory.* ▶ [3]

Geologist J. E. O'Rourke, writing in the *American Journal of Science*, also offers a candid perspective of the reality of circular reasoning practiced by many evolutionary geologists:

◀ *The intelligent layman has long suspected circular reasoning in the use of rocks to date fossils and fossils to date rocks. The geologist has never bothered to think of a good reply, feeling the explanations are not worth the trouble as long as the work brings results. This is supposed to be hard-headed pragmatism.* ▶ [4]

Scientific Reason 37 Summary: The dating of geologic layers and the fossils within them often results from circular reasoning rather than from real scientific observation and testing. Thus, the geologic dates derived from circular reasoning fail to offer credible scientific support to evolution and its proposed billions of years.

⌘ Scientific Reason 38 ⌘ Flat and Horizontal Sedimentary Layers

In many places around the world, we see exceptionally flat horizontal sedimentary sand and rock layers. We can see these layers in the walls of canyons, on the side of a hill that has been dug away, or sometimes along the bank of a roadcut. These consistently flat and horizontal sedimentary layers are often promoted as examples of millions or billions of years of time. But a problem in logic occurs here: if the sedimentary layers were really millions or billions of years old, they would not be exceptionally flat, but rather they would be jagged and uneven.

Here is a hypothetical example to drive home this point about sedimentary layers that were built on land... Picture a large square-mile field prepared and leveled perfectly flat by a bulldozer. This huge field is then left alone for years, say even a 1,000 years, a mere drop in the evolutionary bucket. What would this extremely flat field look like after the 1,000 years? Would it still be very flat and horizontally level, or would it have all kinds of holes, divots, gorges, gulleys, and uneven surfaces on it? Of course, due to rain, flowing water, storms, hail, wind, burroughing and digging animals, animals running or walking across an exposed layer, earthquakes, and plants growing roots, it would not be flat at all! If we magnify this hypothetical scenario to better fit a hypothetical evolutionary timeframe, imagine how uneven our field would be in "only" one million years due to the impact of weather, water, animals, and plants.

A similar point can be made about sedimentary layers that formed in water... After each layer was formed under water and then left alone for many years, due to currents, tidal movements, springs welling up, earthquakes, plants growing into layers, creatures burroughing into layers, and creatures moving across the surface of layers, it too would not remain consistently flat and horizontal after the passing of hundreds or thousands of years.

Thus, observing a series of extremely flat and horizontal layers fails to support long ages; in fact, just the opposite is true. Logically, these extremely flat and horizontal layers actually support very short timeframes and massive amounts of catastrophic, rapid sedimentary layering before anything could disrupt, gouge into, or rip apart the flat and level surface of each layer.

Scientific Reason 38 Summary: The scientific observation that sedimentary rock layers found all over the earth appear very flat and horizontal contradicts what should be observed according to evolution's billions of years. Logically, if these sedimentary layers were really tens or hundreds of millions of years old, due to rain, flowing water, currents, tides, springs welling up, storms, hail, wind, burroughing and digging animals, animals moving across the layers, earthquakes, and plants growing roots, the layers would be full of holes, divots, gorges, gulleys, canyons and uneven surfaces.

⌘ Scientific Reason 39 ⌘
Mount Saint Helens

When Mount Saint Helens first erupted in 1980, the power of the explosion was equivalent to 33,000 Hiroshima-sized atomic bombs! Swirling gases from the explosion reached temperatures of 1,000 degrees Celsius, a blast flume rocketed 15 miles into the atmosphere, 150 square miles of forest was leveled in about 6 minutes, and an estimated 520 million tons of ash billowed eastward for hundreds of miles, with enough ash staying in our atmosphere to slightly decrease the average earth temperature over the next few months. Truly this was a significant geologic event! [5]

In addition to the powerful impact the Mount Saint Helens explosions imposed on life in the surrounding regions, the geologic changes which resulted from the explosions offer surprising support that long ages are not required to explain the geologic sites we see around the world today. Huge sediment deposits, deep canyons, layers of once-living materials starting to become peat and coal, and petrified logs all resulted extremely rapidly. We are not talking millions or billions of years here, but merely a matter of hours, days or years!

Here are five geologic changes that resulted from the multiple Mount Saint Helens explosions during the 1980s: [6]

- The initial explosion leveled 150 square miles of forests near the mountain in about six minutes and left a mat of four million logs floating on what remained of Spirit Lake. The logs were stripped of their bark, some of which settled to the bottom of the lake and created a layer of peat similar to the peat found in many coal deposits. Many of the logs that settled to the lake bottom have remained in a vertical position. Diving expeditions report that today, merely a few decades later, the logs appear similar to a petrified forest.

- The ash and mud flows created massive groups of sediment layers, one of which measures 600 feet thick. Other sediment layers were created out of very fine particles - one such group displays 1,000 miniature layers all within a 25 foot distance. And other sediment layers resemble layers found in canyon walls around the world, such as the Grand Canyon. Amazingly, all of these different groups of layers were created in a matter of hours!

- Massive catastrophic erosion occurred from extensive mudflows and powerful waves of rushing snow-melt water, leaving behind canyons that were formed in less than a day. The resulting Loowit Canyon measures over 100 feet deep and 15 miles long. Step Canyon is even bigger, measuring at some places over 600 feet deep and earning the nickname "Little Grand Canyon".

- Freshly formed volcanic rocks called dacites were left behind after the eruptions. Interestingly, Potassium-Argon radiometric dating performed on the newly produced dacites gave an age of 2.4 million years.

- Life returned to the surrounding region rapidly as fish, animals, and plants reappeared in a matter of years. While many scientists thought the region would stay dormant to life for hundreds if not thousands of years, in actuality, just the opposite occurred. The infusion of ash into the soil served as a fertilizer, and today, merely three-plus decades later, life has been restored to its pre-blast levels.

Scientific Reason 39 Summary: The observable scientific and geologic changes from the eruptions of Mount Saint Helens show that millions of years are not needed to accumulate sedimentary layers, gouge canyons, petrify logs, and form coal layers, all of which evolution proposes took millions of years to form.

⌘ Scientific Reason 40 ⌘
Oil Creation and Pressure

A foundational belief inherent within evolution holds that oil and natural gas resources in the earth's crust are the result of millions or billions of years of time. Actually, we now know that oil and natural gas can be formed in only a matter of a few years or decades.

Geologist and creationist Andrew Snelling, Ph.D., shares that oil and gas can be formed in short time-spans:

◀ *Many people today, including scientists, have the idea that oil and natural gas must take a long time to form, even millions of years. Such is the strong mental bias that has been generated by the prevailing evolutionary mindset of the scientific community. However, laboratory research has shown that petroleum hydrocarbons (oil and gas) can be made from natural materials in short time-spans.* ▶ [7]

Dr. Snelling documents a fascinating experiment performed between 1977 and 1983 at the Commonwealth Scientific and Industrial Research Organization (CSIRO) lab in Sydney, Australia. In summary, during this six-year experiment, different forms of crude oil and natural gas were produced, and brown coal was converted to anthracite, the highest grade of black coal.

Interestingly, not everyone was pleased to hear about the results of this experiment. One type of criticism discredited the results saying they were performed in a man-made lab and, therefore, somehow were not scientifically significant. However, shortly after the results of the CSIRO experiment were published in 1982, the *New York Times* reported that oil was being formed naturally off the coast of California, and it was taking only years to produce this oil. The report states:

◀ *The oil is being formed from the unusually rapid breakdown of organic debris by extraordinarily extensive heat flowing through the sediments, offering scientists a singular opportunity to see how petroleum is formed... Ordinarily oil has been thought to form over millions of years whereas in this instance the process is probably occurring in thousands of years... The activity is not only manufacturing petroleum at relatively high speed but also, by application of volcanic heat, breaking it down into the constituents of gasoline and other petroleum products as in a refinery.* ▶ [7]

Note the significant points from this report:

- Oil forming naturally needs only a few years rather than millions of years.
- The volcanic heat within the earth's crust contributes to these fast rates of oil production.
- The breaking down of oil into gasoline and other petroleum products parallels what an oil refinery would produce.

Geologists also observe that nearly all oil reservoirs display very strong pressure levels. Most measurements of oil reservoir pressure levels give readings of about 20,000 PSI. Due to the known porous structure of much of the earth's crust, and because a liquid under pressure will seek to escape from that pressure, if the earth was really billions of years old, then the majority of oil existing in large pockets within the earth's crust would have already dispersed into the surrounding pores. Thus, it is scientifically illogical to assume that the oil reservoirs in the earth's crust are billions of years old.

Scientific Reason 40 Summary: The scientific observations that oil formation can occur in a few years or decades rather than billions of years and that oil reservoirs have significant pressure which would disperse before billions of years could pass provides scientific support that contradicts evolution's premise of billions of years.

⌘ Scientific Reason 41 ⌘
The Grand Canyon

The Grand Canyon displays awesome geologic wonder. Anyone who sees it walks away inspired by its 277 mile length, its 18 mile width, and its mile depth!

One sedimentary layer partially visible within the side of the Grand Canyon, named the Coconino Sandstone Formation, measures 300 feet thick and 97,000 square miles in area. It displays patterns of water-formed ripple waves over much of its surface, a feature geologists call "large scale cross bedding". Research shows that the Coconino Sandstone Formation developed as a huge underwater sand dune deposit. Based on how water creates patterns in submerged sand, we can deduce that it formed rapidly and catastrophically, not over tens or hundreds of millions of years. [8]

Another notable formation, the Kaibab Upwarp, lies on the north side of the Grand Canyon. It covers 1,150 square miles and ranges from an elevation of 6,000 to 9,200 feet. Within portions of this huge plateau, sections of the layer are actually curved or folded, yet none of these folded portions are cracked or broken. Geologist and creationist Roger Patterson summarizes what we can learn from observing the Kaibab Upwarp, highlighting that the Kaibab Upwarp had to be folded quickly while it was saturated with water:

◀ *Layers that are all similarly bent are strong evidence in support of the rapid deposition of the layers exposed in Grand Canyon. Though there is an alleged time difference of 300 million years between the deposition of the Tapeats Sandstone and of the Kaibab Limestone, the layers have been bent to a similar angle by the Kaibab Upwarp (which allegedly occurred 70 million years ago). In places, the layers of the Tapeats Sandstone are bent at a 90° angle. These formations indicate that the layers must have been soft when the folding occurred. Had they been solid rock they would have fractured, but there is no evidence of fracturing. Heat and pressure can also cause deformation of rocks, but there is no evidence of that in the minerals and structure of these rocks.* ▶ [9]

Both the Coconino Sandstone Formation and the Kaibab Upwarp provide observable evidence that these and similar formations could not have formed through a slow, gradual process needing tens or hundreds of millions of years. For bending to be part of a now-hardened sedimentary layer, the bent layer had to be shaped quickly while it was still very wet. If these layers had been hardened when an attempt to fold them took place, they would have broken or fractured, yet there is no evidence of any fracturing within any of these bent layers.

The Grand Canyon also creates a significant challenge to the concept of the geologic column. (See Scientific Reason 36) Several of the 13 major geological layers composing the geologic column are missing in the Grand Canyon. If the Grand Canyon walls actually displayed a picture of the slow and gradual accumulation of sedimentary layers over millions or billions of years of earth's history, then logically, even accounting for some erosion impact to the layers, all the layers would exist in sequence. Check out the Grand Canyon resources included in the chapter resources for this reason for more in-depth scientific research on this topic.

Scientific Reason 41 Summary: Within the Grand Canyon, both the Coconino Sandstone Formation and the Kaibab Upwarp provide geologic examples that had to be formed from rapid and catastrophic water erosion, not from slow and gradual forces. Within the Grand Canyon, there are layers missing from the proposed geologic column. All of these scientific observations are contrary to evolutionary-based expectations that the Grand Canyon was formed over hundreds of millions of years.

Chapter 5 Endnotes

1 – John Woodmorappe, "The essential nonexistence of the evolutionary-uniformitarian geologic column: a quantitative assessment", *Creation Research Society Quarterly*, Vol. 18 (1981) pages 46-71.

2 – R. H. Rastall, "Geology", *Encyclopedia Britannica*, Vol. 10 (1949) page 168.

3 – Ronald R. West, "Paleoecology and uniformitarianism", *Compass*, Vol. 45 (May 1968) page 216.

4 – J. E. O'Rourke, "Pragmatism versus Materialism in Stratigraphy", *American Journal of Science*, Vol. 276 (January 1976) page 47.

5 – Ken Ham, "I got excited at Mount St Helens", *Creation*, Vol. 15, Iss. 3 (June 1993) pages 14-19. <http://creation.com/i-got-excited-at-mount-st-helens>

6 – John D. Morris, *Creation Basics and Beyond*, Editor Jason Lisle (Dallas, TX: Institute For Creation Research, 2013) pages 223-224.

7 – Andrew A. Snelling, "How fast can oil form?", *Creation*, Vol. 12, Iss. 2 (March 1990) pages 30-34. <http://creation.com/how-fast-can-oil-form>

8 – John H. Whitmore, Raymond Strom, Stephen Cheung, Paul A. Garner, "The Petrology of the Coconino Sandstone (Permian), Arizona, USA", *Answers Research Journal*, Vol. 7 (2014) pages 499-532. <https://answersingenesis.org/geology/rock-layers/petrology-of-the-coconino-sandstone>

9 – Roger Patterson, *Evolution Exposed* (Petersburg, KY: Answers in Genesis, 2008) Chapter 6. <https://answersingenesis.org/geology/geologic-time-scale/geologic-column>

Chapter 5 Resources

Geologic Column

- http://creation.com/geology-questions-and-answers
- http://creation.com/the-collapse-of-geologic-time
- http://creation.com/does-geologic-column-exist
- http://creation.com/geologic-column-global-sequence
- http://www.icr.org/article/242/107
- http://www.icr.org/article/ten-misconceptions-about-geologic-column
- https://answersingenesis.org/geology/geologic-time-scale/geologic-column
- http://www.nwcreation.net/geologycolumn.html
- http://www.trueorigin.org/geocolumn.php

Dating of Geologic Layers

- http://www.icr.org/article/circular-reasoning-evolutionary-biology
- https://answersingenesis.org/theory-of-evolution/molecular-clock/circular-reasoning-surrounds-human-origins-but-even-a-broken-clock-is-right-twice-a-day
- *The New Answers Book 2: Chapter 23* by Answers in Genesis
- *Evolution's Achilles Heel: Chapter 5* by Creation Ministries International

Flat and Horizontal Sedimentary Layers

- http://creation.com/flat-gaps
- http://creation.com/continent-wide-sedimentary-strata
- http://creation.com/its-plain-to-see
- https://answersingenesis.org/geology/rock-layers/transcontinental-rock-layers
- https://answersingenesis.org/geology/rock-layers/2-bent-rock-layers
- https://answersingenesis.org/geology/grand-canyon-facts/no-slow-and-gradual-erosion
- http://www.icr.org/article/flat-gaps-between-strata

Mount Saint Helens

- http://creation.com/i-got-excited-at-mount-st-helens
- http://creation.com/geologist-steve-austin
- https://answersingenesis.org/geology/mount-st-helens/mt-st-helens
- http://www.icr.org/article/lessons-from-mount-st-helens
- http://www.mshcreationcenter.org/7ws1
- http://nwcreation.net/mtsthelens.html

Oil Creation and Pressure

- http://creation.com/how-fast-can-oil-form

The Grand Canyon

- https://answersingenesis.org/geology/grand-canyon-facts/the-grand-canyon
- https://answersingenesis.org/geology/grand-canyon-facts/what-carved-grand-canyon
- https://answersingenesis.org/geology/rock-layers/the-case-of-the-missing-geologic-time
- https://answersingenesis.org/geology/rock-layers/petrology-of-the coconino sandstone
- https://answersingenesis.org/geology/geologic-time-scale/geologic-column
- http://creation.com/startling-evidence-for-noahs-flood
- http://creation.com/geologic-column-global-sequence
- http://creation.com/flat-gaps
- http://www.creationresearch.org/members-only/crsq/42/42_3/2005v42n3p163.pdf
- http://www.icr.org/article/coconino-sandstone-flood-or-desert
- http://www.icr.org/article/transcontinental-sedimentation-flood

⌘ Chapter 6 ⌘

Radiometric Dating

The troubles of the radiocarbon dating method are undeniably deep and serious. Despite 35 years of technological refinement and better understanding, the underlying assumptions have been strongly challenged, and warnings are out that radiocarbon may soon find itself in a crisis situation. Continuing use of the method depends on a fix-it-as-we-go approach, allowing for contamination here, fractionation there, and calibration whenever possible. It should be no surprise then, that fully half of the dates are rejected. The wonder is, surely, that the remaining half has come to be accepted... [3]

⌘ Scientific Reason 42 ⌘ Radiometric Dating Credibility

In order for radiometric dating to be trusted as a scientifically reliable system, it must provide consistent and accurate results. But scientific research has demonstrated many inconsistencies and inaccuracies within radiometric dating, which challenges its scientific credibility.

Described simply, radiometric dating involves running tests on the amounts of radioactive elements within a sample material. For once-living materials like plants and animals, carbon-14 (also known as radiocarbon) dating is performed. For rocks, radioactive dating involves methods such as Potassium-Argon, Rubidium-Strontium, Samarium-Neodymium, and Uranium-Lead dating. For all types of radioactive dating techniques, the testing involves measuring the amount of a particular element within the sample being tested, and by knowing the current decay rate of that specific element, the age of the sample material can be deduced.

Although many evolutionists will speak about radiometric dating as conclusive proof for evolution over billions of years, geologists have discovered that radiometric data often provides inaccurate and inconsistent ages. For example, rock samples from the lava dome created at Mount Saint Helens in 1980 were tested with the Potassium-Argon dating method. The results dated the rock samples between 350,000 and 2,800,000 years old. Such results are of course impossible, because we know exactly when the rocks were formed. Geologist and creationist Tas Walker, Ph.D., highlights these obvious scientific inconsistencies:

◀ *Rock samples from a lava dome within the Mount St Helens crater, USA, were dated using the potassium-argon method. Whole-rock samples gave an age of 350,000 years. When some of the amphibole minerals in the rock sample were extracted and analyzed separately, their age was more than double at 900,000 years. Two mineral samples of a different mineral, pyroxene, gave an age of 1,700,000 and 2,800,000 years. Which age is right? None, actually. The lava dome formed after Mount St Helens exploded in 1980 and the samples were just 10 years old.* ▶ [1]

Another example of radioactive dating inaccuracy occurred with radiometric testing of lava flows from the 1949, 1954, and 1975 volcanic eruptions at Mount Ngauruhoe in New Zealand. The resulting Potassium-Argon dating ages gave timeframes for the lava rock ranging from 270,000 to 3.5 million years ago. [2]

Former editor of the journal *Anthropological Journal of Canada*, Robert E. Lee, highlights the problems with radiometric dating:

◀ *Why do geologists and archeologists still spend their scarce money on costly radiocarbon determinations? They do so because occasional dates appear to be useful. While the method cannot be counted on to give good, unequivocal results, the numbers do impress people, and save them the trouble of thinking excessively. Expressed in what look like precise calendar years, figures seem somehow better... 'Absolute' dates determined by a laboratory carry a lot of weight, and are extremely helpful in bolstering weak arguments... The troubles of the radiocarbon dating method are undeniably deep and serious. Despite 35 years of technological refinement and better understanding, the underlying assumptions have been strongly challenged, and warnings are out that radiocarbon may soon find itself in a crisis situation. Continuing use of the method depends on a fix-it-as-we-go approach, allowing for contamination here, fractionation there, and calibration whenever possible. It should be no surprise then, that fully half of the dates are rejected. The wonder is, surely, that the remaining half has come to be accepted... No matter how 'useful' it is, though, the radiocarbon method is still not capable of yielding accurate and reliable results. There are gross discrepancies, the chronology is uneven and relative, and the accepted dates are actually selected dates. This whole blessed thing is nothing but 13th-century alchemy, and it all depends upon which funny paper you read.* ▶ [3]

Logically, if geologists cannot always accurately date rocks and material of known ages, the credibility of radiometric dating collapses, and it cannot be considered scientifically accurate. Not only should we consider radiometric dating scientifically untrustworthy, many times the dates provided by radiometric testing are thrown out and never reported to the public. Evolutionist J. Ogden III shares:

◀ *It may come as a shock to some, but fewer than 50 percent of the radiocarbon dates from geological and archaeological samples in northeastern North America have been adopted as 'acceptable' by investigators.* ▶ [4]

Additionally, we now know that scientific research has confirmed that radioactive decay rates can change when pressure levels, temperature levels, electric field strengths, and magnetic field strengths are altered. Geologist and evolutionist H. C. Dudley released research that confirmed changes in the decay rates of uranium and thorium. If these radioactive elements which are used to date materials have not always had consistent half-lives and decay rates, then the results from their radioactive dating testing cannot be considered always accurate, which of course strikes at the credibility of any dating results based on uranium and thorium decay rates. [5]

At a Symposium on Radiocarbon Variations and Absolute Chronology held at the University of Uppsala in 1969, scientists Torgny Save-Soderbergh and I. U. Olsson began their presentation by sharing these comments:

◀ *C-14 dating was being discussed at a symposium on the prehistory of the Nile Valley. A famous American colleague, Professor Brew, briefly summarized a common attitude among archaeologists towards it, as follows: If a C-14 date supports our theories, we put it in the main text. If it does not entirely contradict them, we put it in a footnote. And if it is completely out of date we just drop it. Few archaeologists who have concerned themselves with absolute chronology are innocent of having sometimes applied this method.* ▶ [6]

Scientific Reason 42 Summary: Because scientific research has shown that radiometric dating cannot accurately date rocks and materials of known age, that radioactive decay rates can be altered with changes in pressure and temperature, and about half of all resulting dates are ignored or never publically announced, radiometric dating cannot be considered a scientifically reliable dating method, and therefore cannot be used to conclusively support evolution and its billions of years of earth's history.

⌘ Scientific Reason 43 ⌘ Carbon-14 Radiometric Dating

The carbon-14 radiometric dating method measures the amount of carbon-14 in a once-living material, such as coal, wood, bone, shell, and fossil. Living things normally absorb carbon from the atmosphere or their food, but once the plant or animal dies, it no longer absorbs carbon-14. Geologists have been able to determine the current decay rate of carbon 14, and knowing this, the age of once-living material can be deduced.

Carbon-14 breaks down at a current rate that is extremely quick for an assumed evolutionary timeframe of billions of years. The rate that carbon-14 breaks down is known as its "half-life". In a once-living organism, half of the carbon-14 will convert back to nitrogen-14 in 5,730 years. So in two half-lives (11,460 years), 75% of the carbon-14 will have decayed. Extrapolating these amounts and timeframes, any object that is more than 100,000 years old will have no detectable carbon-14 remaining, even when tested with advanced testing equipment called an Accelerator Mass Spectrometer (AMS). This is why carbon-14 dating cannot accurately give results that are millions of years, but instead, just the opposite becomes true: if any material contains any amount of carbon-14, this provides clear scientific support that its maximum age cannot be more than 100,000 years old.

A critical point surfaces here... both evolutionists and creationists have found carbon-14 within fossils, coal, oil, natural gas, limestone, marble, and graphite from every sedimentary layer of the proposed geologic column. Logically, if any of these substances were really millions or billions of years old, they would not have any carbon-14 in them. Scientists have even found carbon-14 in diamonds, so again, diamonds also do not need millions or billions of years to form, and in fact, because they contain carbon-14, they *cannot* be millions or billions of years old. [7]

A team of creationists, all with Ph.D.'s in their respective scientific fields, performed an eight year research project known as Radioisotopes and the Age of the Earth (RATE). After performing radiometric dating research, this team of scientists discovered incredible physical evidence that the earth cannot be millions or billions of years old. Physicist and creationist Don DeYoung, Ph.D., commented on the research performed by the RATE team:

◀ *The RATE scientists are convinced that the popular idea attributed to geologist Charles Lyell from nearly two centuries ago, 'The present is the key to the past', is simply not valid for an earth history of millions or billions of years. An alternative interpretation of the carbon-14 data is that the earth experienced a global flood catastrophe which laid down most of the rock strata and fossils... Whatever the source of the carbon-14, its presence in nearly every sample tested worldwide is a strong challenge to an ancient age. Carbon-14 data is now firmly on the side of the young-earth view of history.* ▶ [8]

Scientific Reason 43 Summary: Geologists have found that fossils, coal, oil, natural gas, limestone, marble, and graphite from every sedimentary layer of the proposed geologic column contain carbon-14. Because any material that contains carbon-14 cannot scientifically be older than 100,000 years, the sedimentary layers as well as the fossils of once-living plants and animals found within them cannot be millions or billions of years old as promoted by evolution.

⌘ Scientific Reason 44 ⌘ Assumptions in Carbon-14 Dating

Carbon-14 dating has become one of the most notable methods of determining the age of once-living material. Evolutionists often assume that the ages offered by carbon-14 dating are conclusive facts, and they promote these dates as proof that coal, wood, bone, shell, and fossils are tens or hundreds of millions of years old. But there are some simple observable and scientifically illogical mistakes with trusting the results of carbon-14 dating.

In reality, carbon-14 dating fails to be scientifically accurate because it is based upon the following seven unverifiable assumptions:

- If the rate of carbon-14 decay has fluctuated at any time in the earth's past, this will impact the reliability of today's reported dates. Is it possible to know that the rate of carbon-14 decay has always been consistent during the entire earth's history? No; there is no scientific way to know this.

- The initial amount of carbon-14 in a plant or animal when it dies must be known in order to date its age accurately. Is it possible to know the exact amount of carbon-14 that was in a living plant or animal when it died years ago? No; there is no scientific or any other way to know this.

- There is another form of carbon in our biosphere known as carbon-12, which is stable and not breaking down in the same way carbon-14 breaks down. In some carbon dating techniques, scientists calculate the ratio of carbon-14 amounts to carbon-12 amounts in order to determine the starting amount of carbon-14 in a living organism at the point it died. The initial ratio of carbon-14 to carbon-12 must be known in order to date accurately. Is it possible to know that the atmospheric ratio of carbon-14 to carbon-12 has always been constant during the entire earth's history? No; there is no scientific way to know this.

- Volcanoes and other unpredictable natural events release additional carbon-14 into our atmosphere. Is it possible to know if the fossil or material being carbon-14 tested was exposed to any additional excess carbon, beyond the normal amount of atmospheric carbon, at any point during its life? No; there is no scientific way to know this.

- The fluctuation of the sun's cosmic rays reaching the earth's atmosphere impacts the breakdown of carbon-14. Is it possible to know if the sun's solar flare energy levels and the intensity of the sun's cosmic rays reaching our atmosphere have always been at consistent levels during the entire history of the earth? No; there is no scientific way to know this.

- The earth's magnetic field strength impacts the breakdown of carbon-14. Per Scientific Reason 20, we know that the earth's magnetic field strength has fluctuated and not always been consistent during the earth's entire history. Thus, there is no scientifically conclusive way to determine how this has impacted the breakdown of carbon-14 during all of earth's history.

- The depths and temperatures of our oceans impact the rate of carbon-14 breakdown within the earth's biosphere. Is it possible to know that the depths and temperatures of our oceans have always been consistent during all of the earth's past? No; there is no scientific way to know this.

The significance of these seven carbon-14 dating assumptions tell us that carbon-14 dating is not scientifically and conclusively reliable, and thus, carbon-14 dating cannot be used to scientifically and conclusively support a belief that the earth is billions of years old. Chemist and creationist Jonathan Sarfati, Ph.D., notes the impact of assumptions when it comes to scientifically dating once-living materials with carbon-14 dating methods:

◀ *We must remember that the past is not open to the normal processes of experimental science; that is, repeatable experiments in the present. A scientist cannot do experiments on events that happened in the past. Scientists do not measure the age of rocks, they measure isotope concentrations, and these can be measured extremely accurately. However, the 'age' is calculated using assumptions about the past that cannot be proven.* ▶ [9]

Scientific Reason 44 Summary: Based on seven scientifically unverifiable assumptions, carbon-14 dating fails to be a conclusively reliable dating method, and thus, cannot be used to support evolution and an earth age of billions of years.

⌘ Scientific Reason 45 ⌘
Carbon-14 SPR Versus SDR

We now have the ability to scientifically measure the production rate of carbon-14, which is known as its Specific Production Rate (SPR). Today in the earth's biosphere, the carbon-14 Specific Production Rate equals 18.8 new atoms per gram of carbon per minute. We can also scientifically measure the rate at which carbon-14 decays, which is its Specific Decay Rate (SDR). Today, the carbon-14 Specific Decay Rate equals 16.1 atom disintegrations per gram of carbon per minute. [10]

Chemist and evolutionist Willard Libby, Ph.D., is credited with discovering the carbon-14 dating method. According to his research, it would take up to 30,000 years for the ratio of SPR and SDR to achieve a stable equilibrium within our biosphere, which is the point at which the SPR and SDR values equal one another. Today, we know that the ratio of SPR and SDR is not in equilibrium, meaning that the decay rate of carbon-14 in the earth's biosphere has not had time to catch up with the production rate of carbon-14. Logically, this scientific observation does not support an earth age of millions or billions of years, but it does indicate that the earth's age must be less than 30,000 years old. [10]

Scientific Reason 45 Summary: The scientifically observed ratio between the Specific Production Rate of carbon-14 and the Specific Decay Rate of carbon-14 is not in equilibrium within the earth's biosphere, which is the opposite of what should be observed if the earth was billions of years old.

Chapter 6 Endnotes

1 – Tas Walker, “Radioactive dating methods”, *Creation*, Vol. 32, Iss. 4 (October 2010). <http://creation.com/radioactive-dating-anomalies>

2 – Andrew A. Snelling, “The Cause of Anomalous Potassium-Argon ‘Ages’ for Recent Andesite Flows at Mt. Ngauruhoe, New Zealand and the Implications for Potassium-Argon Dating”, presented at the Fourth International Conference on Creationism in Pittsburgh, PA, August 3-8, 1998. Published by Creation Science Fellowship, Inc. (1998) pages 503 and 525. <http://www.icr.org/research/index/researchp_as_r01>

3 – Robert E. Lee, "Radiocarbon: ages in error", *Anthropological Journal of Canada*, Vol. 19, No. 3 (1981) pages 9-29.

4 – J. Ogden III, “The Use and Abuse of Radiocarbon”, *Annals of the New York Academy of Science*, Vol. 288 (1977) pages 167-173.

5 – H. C. Dudley, “Radioactivity Re-Examined”, *Chemical and Engineering News* (1April1975) page 2.

6 – Torgny Save-Soderbergh and I. U. Olsson, Presented at the Symposium on Radiocarbon Variations and Absolute Chronology held at the University of Uppsala (1969).

7 – Andrew A. Snelling, “Carbon-14 in Fossils and Diamonds”, *Answers Magazine*, Answersingenesis.org, Pub 1January2011. Web18 September2015. <https://answersingenesis.org/geology/carbon-14/carbon-14-in-fossils-and-diamonds>

8 – Don DeYoung, *Thousands...Not Billions* (Green Forest, AR: Master Books, 2005) page 61.

9 – David Catchpoole, Jonathan Sarfati, and Carl Wieland, *The Creation Answers Book,* Editor Don Batten (Australia: Creation Book Publishers, 2009) page 73.

10 – Mike Riddle, “Chapter 7: Doesn’t Carbon-14 Dating Disprove the Bible?”, *Answers Book 1*, Answersingenesis.org, Pub 20September2007. Web 18September2015. <https://answersingenesis.org/geology/carbon-14/doesnt-carbon-14-dating-disprove-the-bible>

Chapter 6 Resources

Radiometric Dating Credibility

- 💻 http://creation.com/the-collapse-of-geologic-time
- 💻 http://creation.com/the-way-it-really-is-little-known-facts-about-radiometric-dating
- 💻 http://creation.com/radioactive-dating-anomalies
- 💻 http://www.icr.org/rate
- 💻 http://www.icr.org/article/rethinking-carbon-14-dating-what-does
- 💻 http://www.icr.org/article/carbon-14-evidence-for-recent-global
- 💻 http://www.icr.org/research/index/researchp_as_r01
- 💻 http://www.pathlights.com/ce_encyclopedia/Encyclopedia/06dat5.htm
- 📖 *Radioisotopes and the Age of the Earth* by the Institute For Creation Research
- 📖 *Evolution's Achilles Heel: Chapter 6* by Creation Ministries International

Carbon-14 Radiometric Dating

- 💻 http://creation.com/radiometric-dating-questions-and-answers
- 💻 http://www.icr.org/article/young-earth-creation-flood-14c
- 💻 http://www.icr.org/rate
- 💻 https://answersingenesis.org/geology/carbon-14/carbon-14-in-fossils-and-diamonds
- 💻 http://www.answersingenesis.org/articles/am/v6/n1/carbon-14
- 💻 http://www.answersingenesis.org/assets/pdf/am/v7/n4/ECP-p855-864-Radiocarbon-Dating.pdf
- 💻 http://www.answersingenesis.org/assets/pdf/am/v7/n4/Thousands-Not-Billions-ch3-carbon14.pdf
- 📖 *The New Answers Book 1: Chapter 9* by Answers in Genesis

Assumptions in Carbon-14 Dating

- 💻 https://answersingenesis.org/geology/carbon-14/doesnt-carbon-14-dating-disprove-the-Bible
- 💻 http://creation.com/radiometric-dating-questions-and-answers
- 💻 http://www.icr.org/article/117/262
- 💻 http://www.icr.org/article/doesnt-carbon-dating-prove-earth-old
- 📖 *The Creation Answers Book: Chapter 4* by Creation Ministries International

Carbon-14 SPR Versus SDR

- 💻 https://answersingenesis.org/geology/carbon-14/doesnt-carbon-14-dating-disprove-the-bible

⌘ Chapter 7 ⌘

Living Things

The likelihood of the formation of life from inanimate matter [Spontaneous Generation] is 1 to a number with 40,000 naughts [zeroes] after it... it is big enough to bury Darwin and the whole theory of evolution. [1]

⌘ Scientific Reason 46 ⌘ Spontaneous Generation

Evolution relies on the belief that random, purposeless, materialistic, naturalistic, unintelligent, goalless, death-based, designless, chance processes caused life to form billions of years ago, and these processes, following a succession of continual and small genetic and physical advancements over billions of years, are responsible for all life on earth, including people.

A foundational question thus becomes relevant: so how did life start in the first place? If random, purposeless, materialistic, naturalistic, unintelligent, goalless, death-based, designless, chance processes are all that could be responsible for the beginning of everything, per the definition of evolution, then life had to start accidently from non-life. Said another way, life had to start spontaneously. This belief, which becomes absolutely necessary for evolution to be true, has been coined the "Spontaneous Generation" of life.

The Spontaneous Generation of life has never been observed or shown to be possible in any scientific lab experiment, in any genetic testing facility, nor anywhere in nature or our universe. In reality, just the opposite is true. In the 1800s, scientist Louis Pasteur performed a significant experiment that disproved Spontaneous Generation. During this time, most people thought that living things developed out of non-living materials because they saw the sudden appearance of flies and rodents near garbage. His experiment involved sterilizing a substance and then observing that even over time no living things spontaneously arose to life from within the substance. His significant discovery led to advanced pasteurization processes from which we all benefit today, and his experiment established the scientific truth that the Spontaneous Generation of life from non-life is scientifically impossible.

Astronomer and mathematician Sir Fred Hoyle, Ph.D., who did not support belief in evolution or creation, identifies that life from non-living matter, which is required for evolution to be true, is impossible:

◀ *The likelihood of the formation of life from inanimate matter [Spontaneous Generation] is 1 to a number with 40,000 naughts [zeroes] after it... it is big enough to bury Darwin and the whole theory of evolution.* ▶ [1]

Dr. Hoyle also reminds us that the awesome complexity of life could never have arisen by chance:

◀ *As biochemists discover more and more about the awesome complexity of life, it is apparent that its chances of originating by accident are so minute that they can be completely ruled out. Life cannot have arisen by chance.* ▶ [2]

To give perspective to Dr. Hoyle's two statements, the number of electrons in the entire universe has been estimated to be the number 1 followed by 80 zeros, but the scientific odds of life forming from non-living matter is the number 1 followed by 40,000 zeros!

Physicist and evolutionist Freeman Dyson identifies that the Spontaneous Generation of life from non-life has no scientific evidence at all:

◀ *Concerning the origin of life itself, the watershed between chemistry and biology, the transition between lifeless chemical activity and organized biological metabolism, there is no direct evidence at all. The crucial transition from disorder to order left behind no observable traces.* ▶ [3]

Scientific Reason 46 Summary: We now know that the Spontaneous Generation of life from non-life is scientifically impossible, which presents a completely unsolvable contradiction to evolution.

⌘ Scientific Reason 47 ⌘
Law of Biogenesis

The Law of Biogenesis identifies that life cannot originate from non-life, but instead life only originates from other already-existing life.

No scientific research, no scientific studies, and no scientific experiments have ever disputed or disproven the Law of Biogenesis. Consequently, the Law of Biogenesis has never been falsified. Additionally, the Law of Biogenesis is universal and applies everywhere.

Scientific Reason 47 Summary: By definition, evolution requires that life originally arose from non-life. Because this directly contradicts the Law of Biogenesis, a scientific law that has never been falsified, evolution fails to be scientifically true.

⌘ Scientific Reason 48 ⌘
Creating Proteins

One of the simplest free-living organisms we know about is the bacteria, Mycoplasma. It is so simple that it has only one chromosome and a very simple cell membrane with no cell wall. It also has one of the smallest known genomes of any organism that can self-replicate, consisting of only 470 genes. Each gene codes for one protein, and the average size of these proteins is 350 amino acids. Chemist and creationist Dwain L. Ford, Ph.D., shares:

◀ *The probability of forming, by a random assembly method, one such average-size protein molecule containing the amino acid residues in a required sequence is only one in 10 to the 451st.* ▶ [4]

Hypothetically, if the earth was made of pure carbon, it would contain about $10^{(50)}$ carbon atoms. Of course the earth is not made of only carbon, but to put this into perspective based on the simple bacteria Mycoplasma, this bacteria would need $10^{(451)}$ carbon atoms available simply to meet the odds of making all the amino acids needed to form just one protein molecule! As Dr. Ford continues to note:

◀ *In other words, it would require an amount of carbon about $10^{(401)}$ times the size of the earth in order to achieve the probability of forming one required protein molecule!* ▶ [4]

And remember, the Mycoplasma bacteria would need this many carbon atoms to make the amino acids for only one protein molecule that is coded by only one gene, but this very simple bacteria has 470 genes! Using scientific, statistical, and logical reasoning, it becomes apparent that evolution is not capable of explaining how proteins originated, merely by considering the proteins associated with one of the simplest types of bacteria we know about. And if this reality from a simple bacteria with only 470 genes is not enough to make the point, remember that human DNA has billions of genes!

Biologist and evolutionist Joseph W. Thornton, Ph.D., and geneticist and evolutionist Rob DeSalle, Ph.D., identify the scientific impossibility that proteins were originally created by mutations and natural selection:

◀ *Domain shuffling aside, it remains a mystery how the undirected process of mutation, combined with natural selection, has resulted in the creation of thousands of new proteins with extraordinarily diverse and well-optimized functions. This problem is particularly acute for tightly integrated molecular systems that consist of many interacting parts, such as ligands, receptors, and the downstream regulatory factors with which they interact. In these systems it is not clear how a new function for any protein might be selected for unless the other members of the complex are already present, creating a molecular version of the ancient evolutionary riddle of the chicken and the egg.* ▶ [5]

Scientific Reason 48 Summary: It is scientifically, statistically, and logically impossible that evolution and its random, purpose-less processes could be responsible for the origin and existence of proteins.

⌘ Scientific Reason 49 ⌘
Odds of the Formation of Enzymes

The enzymes within our bodies could not be the result of evolution over billions of years of earth's history.

Enzymes, made from protein chains linking 100 or more amino acids, are instrumental in regulating and causing the correct chemical reactions to occur in our bodies. Without enzymes, life could not function. They are critical to nearly every chemical reaction in our body, playing key roles in DNA replication, the digestion of food, and hundreds of cellular activities. Biologists estimate we have about 25,000 enzymes in our bodies!

Mathematicians recognize that 1 chance in $10^{(50)}$ is so near impossible that it has become known as the threshold of possibility. This means that anything with odds greater than 1 in $10^{(50)}$ is considered impossible. If each of the 25,000 enzymes in our bodies were made from only 100 amino acids, the probability of accidently and randomly forming our 25,000 enzymes equals 1 chance in $10^{(2,825,000)}$! [6] Ponder the absolute impossibility of such odds. Because many of the enzymes in our body are actually made from chains built with hundreds of amino acids, the odds are actually even worse than this probability. But whether we use 1 chance in $10^{(2,825,000)}$ or an even more remote probability, it really does not matter, as these statistical odds are completely beyond possible and confirm that evolution cannot explain the origin and formation of enzymes. Microbiologist and creationist Jean Sloat Morton, Ph.D., identifies:

◀ *It is preposterous to think that even one simple enzyme-protein could occur by chance alone, much less the 10 in glycolysis or the 25,000 in the human body!* ▶ [6]

Mathematician and creationist Mike Riddle identifies four factors that we need to recognize when calculating the odds of forming just one protein: [7]

1. We know of over 300 different types of amino acids, but only 20 are used within our bodies. It is not scientifically or logically reasonable to believe that

evolution correctly used only the right 20 amino acids to make the first proteins, and purposely did not use the other 280 amino acids.

2. Each type of amino acid molecule exists in one of two forms: left-handed or right-handed. Significantly, the proteins found within our bodies are made of only left-handed amino acids. Contrast this with every science experiment which has attempted to build amino acids. In every experiment, scientists have observed that left-handed and right-handed amino acids will rapidly bond with each other regardless of their "handedness". Evolution proposes that in an ancient, unknown primordial soup, somehow left-handed amino acids only bonded with other left-handed amino acids to create the proteins found in living things. Evolution could not have correctly selected, by random chance, only left-handed amino acids to make the proteins found in our bodies.

3. Within our bodies, the chains of left-handed amino acids have to line up in an exact sequence in order to properly build proteins. Evolution could not have correctly created, by random chance, the exact sequence of left-handed amino acids required to make proteins.

4. In a supposed ancient unknown primordial soup, if chains of amino acids tried to chemically bond but did not have the protection of a cell membrane, they would have disintegrated in water (a process known as hydrolysis), or they would have disintegrated in an oxygen-based atmosphere (a process known as oxidation), or they would have disintegrated from ultraviolet sun rays in an oxygen-free atmosphere. Evolution has no solution to this three-pronged catch-22 scientific reality.

Here is a final powerful example of the self-refuting reality that evolution could never build amino acids and enzyme proteins: the enzymes that build the amino acid Histidine themselves contain the amino acid Histidine! It is impossible for evolution, which relies on small, successive chemical changes over millions of years, to explain how it acquired the amino acid Histidine in order to evolve the amino acid Histidine.

Biochemist and evolutionist Franklin Harold, Ph.D., sums up the biochemical impossibility that evolution could create enzymes:

◀ *We must concede that there are presently no detailed Darwinian accounts of the evolution of any biochemical or cellular system, only a variety of wishful speculations.* ▶ [8]

Scientific Reason 49 Summary: It is statistically and scientifically impossible for evolution to have formed, by random chance, the estimated 25,000 enzymes that function in a human body. Evolution cannot circumvent the scientific and statistical impossibility of originally forming an enzyme by random chance.

⌘ Scientific Reason 50 ⌘
Cells

The complexity, design, and beauty displayed by cells and their fully functioning internal systems cannot be the result of evolution over billions of years.

Charles Darwin thought the cell was just a simple blob of liquid, but in reality, a living cell is so incredibly complex that it defies any level of evolutionary explanation. Recent advancements in cell biology have taught us that a cell can contain over 60,000 proteins, each built using one of millions of different amino acid configurations. Cells have complex cell walls, an energy processing system, a system of self-repair, a reproductive system, a method for gaining energy and expelling cell waste, and methods for recognizing and obeying the DNA instructions found within their nucleus. The body of an adult person, depending on their height and size, consists of 17 trillion cells. To put this in perspective, think about this powerful visual: If all the cells from a single adult person were lined up side by side, they would circle the earth many times.

Evolution proposes that the first cell formed by random and directionless processes billions of years ago in a supposed primordial soup. A reason we know this cannot be true comes from the fatty acids and phosphate molecules that are needed to build cell parts. Fatty acids are needed to make cell walls and phosphate molecules are necessary to build DNA, RNA, ATP, and other required molecules within cells. But all available geologic research tells us that our oceans have always included calcium ions, and these calcium ions would cause the supposed freely-floating fatty acids and phosphate molecules to clump (chemically bond) together in any supposed primordial soup. Thus the fatty acids and phosphates would be unavailable for use in any supposed chemical evolution of cells, providing chemistry-based logic that it is impossible for cells to have formed in any oceanic primordial soup.

Biochemist and evolutionist David Ezra Green, Ph.D., along with medical doctor, molecular biophysicist and evolutionist Robert F. Goldberger, M.D., identify how scientifically impossible it is to consider evolution responsible for creating cells:

◀ *The popular conception of primitive cells as the starting point for the origin of the species is really erroneous. There was nothing functionally primitive about such cells. They contained basically the same biochemical equipment as do their modern counterparts. How, then, did the precursor cell arise? The only unequivocal rejoinder to this question is that we do not know.*

Although seven steps are shown, leading from atoms to ecosystems, there is one step that far outweighs the others in enormity: the step from macromolecules to cells. All the others can be accounted for on theoretical grounds - if not correctly, at least elegantly.

However, the macromolecule-to-cell transition is a jump of fantastic dimensions, which lies beyond the range of testable hypothesis. In this area all is conjecture. The available facts do not provide a basis for postulating that cells arose on this planet. ▶ [9]

The complex process of cell division cannot be explained by evolution. Our cells divide and produce two daughter cells, with each daughter cell having exact replicas of their parent DNA and cell organelles. Our cells do this both for growth and repair of our tissues. Of the estimated 17 trillion cells which comprise an adult person, 80 billion are replaced each day! Except for red blood cells which lack DNA, this means the six billion DNA nucleotides within each cell have to be correctly replicated in the right sequence as well. Geneticist and creationist Jeffrey Tomkins, Ph.D., highlights the significance that evolution cannot explain how a cell could evolve:

◀ *The question of how life first arose on Earth is perhaps the greatest obstacle for the evolutionary paradigm. While the whole concept of biological evolution itself is full of serious problems, the origins of the first biomolecules and the first cell, not to mention the enormous amount of information contained within the cell, is a complete impossibility from a naturalistic perspective. In fact, without a plausible explanation as to the origin of the first cell, the whole evolutionary story collapses.* ▶ [10]

Microbiologists have also discovered that a cell provides another example of irreducible complexity because all of its parts must exist and fully work together or a cell would not exist. Astrophysicist and creationist Jason Lisle, Ph.D., summarizes this reality:

◀ *Scientists have found that within the cell there are thousands of what can be called 'biochemical machines.' For example, one could cite the cell's ability to sense light and turn it into electrical impulses. But what scientists once thought was a simple process within a cell, such as being able to sense light and turn it into electrical impulses, is in fact a highly complicated event. For just this one example alone to work, numerous compounds must all be in the right place, at the right time, in the right concentration - or it just won't happen. In other words, just as all the parts of a 747 need to be assembled before it can fly, so all the parts of these 'biochemical machines' in cells need to be in place, or they can't function. And there are literally thousands of such 'machines' in a single cell that are vital for it to operate.* ▶ [11]

Biochemist and intelligent design advocate Michael Behe, Ph.D., notes the irreducible complexity (See Scientific Reason 55) displayed within cells could not be the result of evolution:

◀ *Now it's the turn of the fundamental science of life, modern biochemistry, to disturb. The simplicity that was once expected to be the foundation of life has proven to be a phantom; instead, systems of horrendous, irreducible complexity inhabit the cell. The resulting realization that life was designed by an intelligence is a shock to us in the twentieth century who have gotten used to thinking of life as the result of simple natural laws. But other centuries have had their shocks, and there is no reason to suppose that we should escape them.* ▶ [12]

Scientific Reason 50 Summary: It is biologically, chemically, and statistically impossible for the cells in our body, which include DNA, RNA, proteins, cellular organelles and membranes, to be the result of random, spontaneous chance-based processes. Therefore it is scientifically impossible for a supposed first cell to have formed according to a series of small, successive genetic and physical changes over billions of years as promoted by evolution.

⌘ Scientific Reason 51 ⌘
Bacterium Flagellum

The amazing structure, design and complexity of the bacterium flagellum could never be the result of random and unintelligent processes promoted by evolution.

A flagellum is a tail-like structure used by bacteria to propel themselves through liquid. It contains a rotary engine made from 40 different proteins and generates power by the movement of protons through the center of its rotary motor. A bacterium can rotate its flagellum clockwise or counter-clockwise, and this circular rotation propels a bacterium through liquid at a speed of 60 cell lengths per second. To put this into perspective, this is equivalent to a six-foot tall man running over 100 yards in a second! [13]

The motor that rotates the flagellum displays amazing design, especially considering its microscopic size. The motor includes a rotor, a stator, a clutch, multiple layers within the inner and outer membranes, bearings, bushings, a hook, and a flagellar filament. To appreciate how small a flagellum and its motor actually is, microbiologists estimate its size at about 1 millionth of an inch. Not only does its tiny size cry out as an example of purpose and design, the motor's rotor spins at speeds up to 100,000 revolutions per minute! [13]

Thousands of evolutionary scientists have studied and researched the structure and design of the bacterium flagellum motor, yet none have ever successfully proposed a model that can scientifically account for its existence and still be consistent within the limitations of evolutionary beliefs.

Scientific Reason 51 Summary: The incredible structure, design, and complexity of a bacterium's flagellum screams of design and purpose that cannot be the result of the random chance-based processes proposed by billions of years of evolution.

⌘ Scientific Reason 52 ⌘
Butterfly Metamorphosis

The incredible process of butterfly metamorphosis, during which a caterpillar first changes into a chrysalis and then into a butterfly, cannot be explained by evolution over millions of years.

Butterfly metamorphosis first involves a caterpillar taking three days to change into a cocoon-like structure called a chrysalis. This chrysalis contains a liquid green goo, DNA, and a beating heart. Then, about nine days later, a beautifully designed and flight-capable butterfly emerges from the chrysalis.

There is no possible way to explain how evolution could cause the three stage process of butterfly metamorphosis to occur quickly, accurately and in the proper sequence. Evolution promotes the idea that millions of small, successive physical changes in a creature, due to billions of genetic mutations creating brand new DNA, occur as responses to environmental factors and accumulate over millions of years of time. Directly contradicting the timeframe proposed by evolution, the butterfly changes from a caterpillar to a small chrysalis in only three days, and then from a chrysalis to a fully functional, flight-capable butterfly in another nine days. It is scientifically and observationally obvious that there is just not enough time for evolution to even occur.

Additionally, the process of changing from a caterpillar to a chrysalis containing only green goo, DNA, and a beating heart contradicts what should be observed if evolution was true. Per evolutionary beliefs, creatures are always improving and gaining brand new DNA and body parts, but a caterpillar morphing into a simplistic green goo with only DNA and a beating heart is a massive reduction in body complexity and form, certainly not any form of improvement!

Richard Milton, who does not support belief in evolution or creation, highlights that science cannot explain the process of butterfly metamorphosis:

◀ *...no stage or aspect of this physical process can be accounted for or even guessed at with our current knowledge of chemistry, physics, genetics, or molecular biology, extensive though they are. It is completely beyond us. We know practically nothing about the plan or program governing the metamorphosis, or the organizing agency that executes this plan.* ▶ [14]

Evolutionary timescales do a nosedive when attempting to explain how a caterpillar can change into a chrysalis and then change into a butterfly in less than two weeks. Additionally, evolution cannot explain the butterfly's amazing abilities and features. Butterfly wings, when viewed by an electron microscope, display hundreds of rows of intricately designed and patterned structures which refract sunlight and shimmer in brilliant and fluorescent colors. Scientific research has taught us that butterflies navigate thousands of miles using the earth's magnetic field and sunlight. Butterflies have complex eyes that are sensitive to the sun's polarized light, with each eye housing 6,000 separate lenses giving the butterfly 360 degree vision!

Scientific Reason 52 Summary: It is scientifically impossible for the fascinating, two-week process of butterfly metamorphosis as well as the butterfly's amazing abilities to be the result of evolution over millions of years.

⌘ Scientific Reason 53 ⌘
Animal Design, Beauty, and Amazing Capabilities

Many animals display incredible design, beauty, and capabilities that cannot be explained by evolution. Evolution proposes that millions of small, successive physical changes have occurred in animals due to genetic mutations correctly creating billions of brand new genes in response to environmental factors. Here are nine inspiring animal capabilities that could never result from evolution's random chance-based processes:

- The Artic Tern travels a 12,000 mile journey between the Arctic and the Antarctic! Ornithologists believe that the Artic Tern has the ability to navigate by recognizing wind pathways, the magnetic field of the earth, the location of the sun and stars, the flow of ocean currents, and ocean plant levels. These fully-functioning advanced navigational abilities cannot be explained as results of random and unintelligent processes offered by evolution.

- The marvelous insect Issus Coleoptratus lives on European garden leaves. Entomologists have discovered that the baby nymph has intermeshing gears within its body, which cause its back legs to simultaneously push within 30 microseconds of each other. These incredible gears consist of two circular sets of "toothed wheels" that are each 30 micrometers tall and 400 micrometers in length. The Issus Coleoptratus' fully functioning gears could never be scientifically explained by small, successive, random genetic and physical changes offered by evolution, yet this insect could not hop correctly without these gears. [15]

- A male Bombyx Mori beetle can detect a single molecule of the female's pheromone bombykol with its antennas, which are each lined with 17,000 sensory hairs. Note that just one cubic centimeter of air contains 27 million billion molecules! The random chance-based processes inherent within evolution cannot explain this amazing sensing ability. [16]

- The incredible precision of a dolphin's sonar system enables it to detect a 2 inch object up to 230 feet away, and marine biologists have discovered its sonar click pattern is mathematically calibrated in the most productively possible way. The dolphin's sonar system includes a "melon" of fatty tissue within the top of its head, which functions like a sound funnel to direct sound waves into a beam and then aim the beam. The melon consists of different types of fats arranged in the necessary shapes and sequences to bend ultrasonic sound waves correctly. Amazingly, not only do the sound waves accurately travel to the dolphin's target, but the melon generates these sound waves in the required patterns to cause them to correctly return to the dolphin! Evolution proposes that unintelligent, designless, purely-naturalistic processes led to the formation of every creature, but it is scientifically and logically impossible that unintelligent, designless, purely-naturalistic forces could design the advanced sonar system in a dolphin.

- Some grasshoppers have the ability to detect vibrations which have an amplitude of only .0000000050 centimeters, which is equal to a fraction of the diameter of the first electron orbit shell in a normal hydrogen atom. The random chance-based processes promoted by evolution cannot explain this amazing sensing ability. [17]

- A hummingbird's wings beat an average of 80 times per second, and in a steep dive, its wings can reach an outrageous 150 beats per second. It can stop, move forward, move backward, move down, move up, move diagonally, accelerate, and decelerate in a fraction of a second. The design and structure of its body and wings to enable such incredible acrobatic and aerodynamic abilities could never be the result of the purposeless and designless processes inherent within evolution.

- Marine biologists have discovered that lobster eyes are designed based upon exact geometrical relationships between the parts of its eyes. Astronomers have designed advanced telescopes that mimic the design features found in lobster eyes, which has enabled them to build telescopic lenses capable of focusing x-rays received from distant stars. The random and designless forces proposed by evolution certainly cannot be the source of the lobster's advanced optical technology. [18]

- Vipers have heat sensitive organs on the side of their head used to detect the heat radiation of its prey. These organs interpret wavelengths varying from infrared to ultraviolet, regardless of its own body temperature at the moment. It is not possible for the random chance-based processes offered by evolution to explain these snake's amazing sensing capability. [19]

- The complex eyes of trilobites are an example of incredible design. Their eyes were made of tubes that each pointed in a different direction and had special lenses to focus light from any distance. The lens design used a layer of calcite over a layer of chitin, with both layers designed with just the right light refraction index. Physicists have come to learn that the design of trilobite eyes used aspects of several optical laws of physics, including Fermat's principle of least time, Snell's law of refraction, Abbe's sine law, and some aspects of birefringent optics. Trilobites are often promoted as an alleged primitive index fossil by proponents of evolution, but demonstrating these amazing aspects of optical physics is certainly not primitive! [20]

Scientific Reason 53 Summary: The incredible design, beauty, and capabilities displayed by the 12,000 mile journey of the Artic Tern, the acute sensing ability of the Bombyx Mori beetle, the advanced sonar system in dolphins, the highly sensitive vibration detection system in grasshoppers, the advanced optics in lobster eyes, the heat radiation sensing ability of vipers, and the complex design of Trilobites eyes are all examples of capabilities that the random chance-based processes proposed by evolution over billions of years could never create.

⌘ Scientific Reason 54 ⌘ Patterns of Discontinuity

We observe that all creatures have a recognizable appearance, displaying many characteristics which are easily distinguishable from other creatures. Consequently, all creatures can be organized and classified into logical kinds because of their distinctly recognizable and specific appearance, and these distinctions contradict any sort of evolving continuity that the belief in evolution promotes.

If evolution was true, there would be millions of creatures that appear to be a hybrid of two or more creatures. Sometimes the Platypus is brought up an example of animal continuity, but it truly is an exception in the animal kingdom and not the norm of what we observe. Geologist and creationist Kurt Wise, Ph.D., summarizes the animal discontinuity we actually observe in the animal world:

◀ *Discontinuities, for example, divide the animal kingdom into about three dozen distinct groups called phyla. Consider just a few of the profound differences between phyla. Chordates have skeletons inside their bodies, whereas arthropods have skeletons outside their bodies. Even among animals that protect themselves with very specific armor plates, such as two carbonate shells, different phyla often put them together in radically different ways: clams place the shells on the sides of their bodies, while brachiopods place them on the front and back of their bodies. Sponges are identical, or symmetric, in every direction around a line, while echinoderms, such as starfish, are symmetric in five directions around a line, and worms are symmetric along a plane... Looking deeper, discontinuity can also be found within phyla. There is such discontinuity between turtles and all other reptiles (for instance, turtles have shoulder blades inside their body cavity) and between bats and all other mammals that evolutionary transitions are not only unknown, but sometimes even difficult to imagine.* ▶ [21]

The animal kingdom is categorized into groups called phyla which are based upon a variety of animal characteristics. As just mentioned, even within these phyla, there is no consistent appearance of body parts that would be expected if all the animals had evolved from a common ancestor.

Scientific Reason 54 Summary: The pervasive pattern of discontinuity observed in the appearance and structure of all living things displays exactly the opposite of what should be observed if evolution over billions of years was true.

⌘ Scientific Reason 55 ⌘
Irreducible Complexity

We observe many systems and relationships in living creatures which demonstrate irreducible complexity. Our blood clotting system serves as a fascinating example.

The words "irreducible complexity" can sound fairly complicated, but like most things, when broken down into a simple definition, they become easy to understand. In a nutshell, something demonstrates irreducible complexity when all its necessary interacting parts must both exist and fully work. If any of the needed parts are removed or fail to work, the entire system will not work. So in the example of our blood clotting system, if just one of the chemicals was not available or a clotting step did not work correctly, our blood could not clot when needed.

This reason becomes significant because evolution balances on the idea that all life resulted from millions and millions of small, successive genetic and physical steps building a living creature or system. But in reality, any system that demonstrates irreducible complexity could never develop in this way. Why? Because any early steps or versions of the creature or system would not have the needed fully-functioning parts to operate, and logically, the system would not work. Said another way, one of the fundamental problems with evolution is that it would produce a creature, a system, or an organ that would be a non-functional developing component by the evolutionary process. But that non-functional developing intermediate component or stage would not work or survive and therefore would provide nothing to evolve to the next evolutionary development stage.

Biochemist and intelligent design advocate Michael Behe, Ph.D., summarized that any system that demonstrates irreducible complexity could not be the result of evolution:

◀ *What type of biological system could not be formed by 'numerous, successive, slight modifications'? Well, for starters, a system that is irreducibly complex. By irreducibly complex, I mean a single system composed of several well-matched, interacting parts that contribute to the basic function, wherein the removal of any one of the parts causes the system to effectively cease functioning. An irreducibly complex system cannot be produced directly (that is, by continuously improving the initial function, which continues to work by the same mechanism) by slight, successive modifications of a precursor system, because any precursor to an irreducibly complex system that is missing a part is by definition non-functional.* ▶ [22]

Charles Darwin, the man most recognized for suggesting the idea of evolution, said that if any complex organ exists which demonstrates irreducible complexity, it could not be the result of evolution:

◀ *If it could be demonstrated that any complex organ existed, which could not possibly have been formed by numerous, successive, slight modifications, my theory would absolutely break down.* ▶ [23]

In the years since 1859 when Charles Darwin first proposed his idea of evolution, we have discovered many examples of irreducible complexity that all refute his suggestion. Powerful examples include photosynthesis (See Scientific Reason 56), seeds (See Scientific Reason 57), DNA replication (See Scientific Reason 62), and the human eye (See Scientific Reason 65). The scientific community today has no conclusive scientific explanation for how random, purposeless, chance processes can explain the widespread existence of irreducible complexity within living things.

Scientific Reason 55 Summary: It is impossible for the irreducible complexity demonstrated in living things to be the result of billions of small, continual, successive genetic and physical steps offered by evolution over billions of years.

⌘ Scientific Reason 56 ⌘ Photosynthesis

Photosynthesis is one of the most incredible processes supporting life on earth. A simple definition: it is the process in which green plants (and some bacteria and algae) use sunlight to convert water, carbon dioxide, and minerals into oxygen, amino acids, proteins, fats, pigments, and chemical energy (carbohydrates and sugars) within its cells.

In plants, photosynthesis occurs in cell organelles called Chloroplasts, and these Chloroplasts contain Chlorophyll, the green pigment which absorbs sunlight. A single plant leaf can contain 70 million cells. Together all these cells combine to house five billion Chloroplasts, with each of these individual Chloroplasts containing 600 million molecules of Chlorophyll. Keep in mind that this is for just one leaf, and many plants have tens of thousands of leaves. [24]

Although the process of photosynthesis occurs in both land and water plants, marine biologists estimate that ocean phytoplankton photosynthesis produces 70% of the earth's oxygen. Regardless of where the majority of photosynthesis happens, without it, plants would die, and life on the earth would cease.

The process of photosynthesis involves complicated chemical systems and hundreds of chemical reactions that must occur in a specific sequence. It involves chemical interdependencies between plant parts and molecules of carbon dioxide and oxygen, all which must work together in a correct chemical sequence in order to produce water, amino acids, proteins, fats, pigments, and the production of Adenosine Triphosphate (ATP) molecules. As is commonly true in the field of science, the process of photosynthesis could never occur as a partial or intermediate scenario. As yet another example of irreducible complexity, it is impossible for evolution to cause photosynthesis to occur by random and accidental causes.

Scientific Reason 56 Summary: Because the process of photosynthesis involves complicated chemical systems and hundreds of chemical reactions that must occur in a specific sequence, it is an irreducibly complex chemical system which could never be the result of random and accidental causes proposed by evolution over billions of years of earth's history.

⌘ Scientific Reason 57 ⌘
Seeds

Although a very small and simple object, a seed provides a huge example of design and complexity. The process of seed germination has been studied for years, but botanists still cannot completely explain how and why seeds grow.

Within each seed is information (DNA), a small amount of stored energy, and tiny parts necessary to eventually grow and produce a fully grown plant. A redwood tree reminds us how such a tiny seed can lead to an enormous plant - the biggest redwood tree reaches about 360 feet tall! [25]

In order for a huge redwood tree to grow from its original seed, all the required genetic information, a system for reading and following the genetic information, the required energy storage, a process to use the stored energy, and the required initial seed parts all have to be fully formed and fully functional in order for the seed to germinate and grow into a fully developed plant. If any one of these requirements is missing, the seed is worthless. There could be no such thing as an intermediate version of a seed that was only partially capable of growing into a fully grown plant comprised of trillions and trillions of specifically purposed cells.

Scientific Reason 57 Summary: In seeds, the design, the ability to follow genetic instructions, the ability to use stored energy, and the ability to grow into a fully developed plant with trillions and trillions of specifically purposed cells all demonstrate irreducible complexity, none of which can be scientifically explained by evolution.

⌘ Scientific Reason 58 ⌘
Fermi Paradox

Physicist and evolutionist Enrico Fermi, Ph.D., suggested that if evolution was true throughout our universe, there would be other forms of life on at least one of the other billions of planets in the billions of galaxies within our universe. Logically, he argues, if there was other life in our universe, people on earth would have evidence of this life by now. [26]

But, in spite of all the hype and science fiction promoting aliens and extraterrestrial life forms, there is not one shred of conclusive proof for any alien or extraterrestrial life form, not even a single-celled life form. In short, ET and his friends do not exist! Enrico Fermi recognized this scientific truth, and it has since been coined the "Fermi Paradox".

Scientific Reason 58 Summary: While the reality of the Fermi Paradox certainly does not by itself scientifically remove support for evolution, it does provide one more example of scientific observations which are consistent with the impossibility of evolution over billions of years.

Chapter 7 Endnotes

1 – Lee Elliot Major, "Big enough to bury Darwin", *Guardian (UK) Education Supplement* (23August2001). <http://www.theguardian.com/education/2001/aug/23/highereducation.peopleinscience>

2 – Fred Hoyle, *The Intelligent Universe* (London: Michael Joseph Limited, 1983) page 11.

3 – Freeman Dyson, *Origins of Life* (New York: Cambridge University Press, 1999) page 36.

4 – Dwain L. Ford, *In Six Days – Why 50 Scientists Choose to Believe in Creation*, Editor John F. Ashton (Green Forest, AR: Master Books, 2001) page 139.

5 – Joseph W. Thornton and Rob DeSalle, "Genomics Meets Phylogenetics", *Annual Review of Genomics and Human Genetics* (2000) page 64.

6 – Jean Sloat Morton, "Glycolysis and Alcoholic Fermentation", *Acts & Facts*, Vol. 12 (1980) page 9. <http://www.icr.org/article/glycolysis-alcoholic-fermentation>

7 – Mike Riddle, "Chapter 16: Does Evolution Have a...Chance?", *The New Answers Book 3*, Answersingenesis.org, Pub 30May2014. Web 18September2015. <https://answersingenesis.org/evidence-against-evolution/probability/does-evolution-have-a-chance>

8 – Franklin M. Harold, *The way of the cell: molecules, organisms and the order of life* (New York: Oxford University Press, 2001) page 205.

9 – David Ezra Green and Robert F. Goldberger, *Molecular Insights into the Living Process* (New York: Academic Press, 1967) pages 403 and 406.

10 – Jeffrey Tomkins, *Creation Basics and Beyond*, Editor Jason Lisle (Dallas, TX: Institute For Creation Research, 2013) page 185.

11 – Jason Lisle, *The New Answers Book 1*, Editor Ken Ham (Green Forest, AR: Master Books, 2010) page 16.

12 – Michael J. Behe, *Darwin's Black Box* (New York: The Free Press, 1996) pages 252–253.

13 – Steven DeVowe, "The Amazing motorized germ", *Creation Magazine*, Vol. 27, Iss. 1 (December 2004). <http://creation.com/the-amazing-motorized-germ>

14 – Richard Milton, *Shattering the Myths of Darwinism* (Rochester, VT: Park Street Press, 1992) page 220.

15 – Jonathan Sarfati, "Toothed gears in jumping insects", *Creation.com*, Pub 19September2013. Web 4January2016. <http://creation.com/toothed-gears-jumping-insects >

16 – Takeshi Sakurai, Shigehiro Namiki, Ryohei Kanzaki, "Molecular and neural mechanisms of sex pheromone reception and processing in the silkmoth Bombyx mori", *Frontiers in Physiology*, Journal.Frontiersin.org, Pub 31March2014. Web 4January2016. <http://journal.frontiersin.org/article/10.3389/fphys.2014.00125/full>

17 – Werner Gitt, "What Is Life? The Informational View", *Answers in Genesis*, Answersingenesis.org, Pub 1November2012. Web 4January2016. <https://answersingenesis.org/genetics/information-theory/54-what-is-life-the-informational-view-ob19>

18 – Chris Ashcraft, "New design innovations from biomimetics", *Creation Magazine*, Vol. 32, Iss. 3 (July 2010) pages 21-23. <http://creation.com/lobster-eye-design>

19 – "Viper", *CreationWiki, the Encyclopedia of Creation Science*, Creationwiki.org, Pub 13December2011. Web 4January2016. <http://creationwiki.org/viper>

20 – Charles Stammers, "Incredible lens engineering in an 'early' creature", *Creation Magazine*, Vol. 21, Iss. 1 (December 1998) page 23. <http://creation.com/trilobite-technology>

21 – Kurt Wise, "The Discontinuity of Life", *Answers Magazine*, Answersingenesis.org, Pub 1January2009. Web 18September2015. <https://answersingenesis.org/evidence-for-creation/the-discontinuity-of-life>

22 – Michael J. Behe, *Darwin's Black Box* (New York: The Free Press, 1996) page 39.

23 – Charles Darwin, *On the Origin of Species by Means of Natural Selection, or the Preservation of Favoured Races in the Struggle for Life: A Facsimile of the First Edition* (Cambridge, MA: Harvard University Press, 1964) page 189.

24 – Rick Swindell, "Shining light on the evolution of photosynthesis", *Journal of Creation*, Vol. 17, Iss. 3 (December 2003) pages 74-84. <http://creation.com/shining-light-on-the-evolution-of-photosynthesis>

25 – Gary Bates, "Patriarchs of the Forest", C*reation*, Vol. 25, Iss. 1 (December 2002) pages 10-13. <http://creation.com/patriarchs-of-the-forest>

26 – Russell Grigg, "Into the Universe with Stephen Hawking: Aliens", *Creation Ministries International*, Creation.com, Pub 1December2012. Web 4January2016. <http://creation.com/stephen-hawking-aliens>

Chapter 7 Resources

Spontaneous Generation

- http://creation.com/the-origin-of-life
- https://answersingenesis.org/origin-of-life/the-origin-of-life-a-problem-for-evolution
- http://www.theguardian.com/education/2001/aug/23/highereducation.peopleinscience
- *Evolution the Grand Experiment: Chapter 2* by Carl Werner

Law of Biogenesis

- http://creation.com/origin-of-life
- http://creation.com/life-from-lifeor-not
- http://www.icr.org/article/7911

Creating Proteins

- http://www.annualreviews.org/doi/abs/10.1146/annurev.genom.1.1.41
- http://creation.com/did-god-create-life-ask-a-protein
- http://creation.com/the-origin-of-life
- *In Six Days – Why 50 Scientists Choose to Believe in Creation: Chapter 14* by Master Books
- *Evolution the Grand Experiment: Chapter 18* by Carl Werner

Odds of the Formation of Enzymes

- http://creation.com/self-replicating-enzymes#r4
- http://www.icr.org/article/glycolysis-alcoholic-fermentation
- http://www.icr.org/article/105/358
- http://www.icr.org/article/life-theories-origin-biological-order
- https://answersingenesis.org/origin-of-life/can-natural-processes-explain-the-origin-of-life
- https://answersingenesis.org/evidence-against-evolution/probability/does-evolution-have-a-chance
- *In Six Days – Why 50 Scientists Choose to Believe in Creation* by Master Books

Cells

- http://creation.com/life-is-in-the-blood
- *Evolution: A Theory in Crisis* by Adler & Adler
- *Evolution Is Biologically Impossible* by Institute For Creation Research
- *Creation Basics and Beyond: Chapter 16* by Institute For Creation Research

Bacterium Flagellum

- http://creation.com/the-amazing-motorized-germ
- http://creation.com/germs-miniature-motor-has-a-clutch
- http://www.icr.org/article/molecular-motors-vs-evolutionism
- http://www.icr.org/article/debating-design-bacterial-flagellum

Butterfly Metamorphosis

- http://creation.com/beautiful-black-and-blue-butterflies
- http://creation.com/butterfly-brilliance
- http://creation.com/butterfly-designer-wings
- http://creation.com/the-magnificent-migrating-monarch
- http://www.icr.org/article/butterflies-vs-macroevolution

Animal Design, Beauty, and Amazing Capabilities

- http://creation.com/toothed-gears-jumping-insects
- http://creation.com/lobster-eye-design
- http://creation.com/trilobite-technology
- http://journal.frontiersin.org/article/10.3389/fphys.2014.00125/full
- https://answersingenesis.org/genetics/information-theory/54-what-is-life-the-informational-view-ob19
- http://creationwiki.org/Viper
- *Did God use Evolution* by Master Books
- *In Six Days – Why 50 Scientists Choose to Believe in Creation* by Master Books

Patterns of Discontinuity

- http://creation.com/cladistics
- https://answersingenesis.org/evidence-for-creation/the-discontinuity-of-life

Irreducible Complexity

- http://creation.com/refuting-evolution-2-chapter-10-argument-irreducible-complexity
- http://creation.com/the-design-of-tears-an-example-of-irreducible-complexity
- http://www.icr.org/article/irreducibly-complex-genome-designed
- https://answersingenesis.org/evidence-against-evolution/irreducible-complexity-some-candid-admissions-by-evolutionists

Photosynthesis

- http://creation.com/green-power-photosynthesis
- http://creation.com/c4-photosynthesisevolution-or-design
- http://creation.com/shining-light-on-the-evolution-of-photosynthesis
- https://answersingenesis.org/biology/plants/little-green-machines

Seeds

- 💻 http://creation.com/sensational-seeds
- 💻 https://answersingenesis.org/biology/plants/the-origin-of-plants

Fermi Paradox

- 💻 https://answersingenesis.org/answers/books/taking-back-astronomy/the-bible-and-modern-astronomy-part-2
- 💻 http://creation.com/stephen-hawking-aliens

⌘ Chapter 8 ⌘

Information and DNA

Do we, therefore, ever see mutations going about the business of producing new structures for selection to work on? No nascent organ has ever been observed emerging, though their origin in pre-functional form is basic to evolutionary theory. Some should be visible today, occurring in organisms at various stages up to integration of a functional new system, but we don't see them. There is no sign at all of this kind of radical novelty. Neither observation nor controlled experiment has shown natural selection manipulating mutations so as to produce a new gene, hormone, enzyme system or organ. [11]

⌘ Scientific Reason 59 ⌘ Information

Information cannot arise by accident, but instead, greater information is always required to produce subsequent information. One of our most brilliant scientists, Albert Einstein, spoke of this truth. He explained that matter and energy alone could never give meaning to symbols and language, and thus, information and language must be thought of as a category of reality separate from matter and energy. To this day, his insights have been coined the "Einstein Gulf". [1]

On one side of the Einstein Gulf exist material objects and space-time relationships, and on the other side exist abstract ideas and concepts, including information and language. Communications professor and creationist John W. Oller, Jr., Ph.D., sheds light on the inability of any evolutionary belief to bridge the Einstein Gulf:

◀ *As far as we can tell from science, hard objects never just accidentally happen to cross Einstein's Gulf, and logic shows why such an accidental crossing is impossible. In fact, Einstein said that the 'gulf' was 'logically unbridgeable' (1944, page 289). Neither can the events that involve hard objects, nor the space-time relations that they constitute, just automatically and accidentally turn themselves into comprehensive propositions or prepositional relations, such as are found in a true narrative (of the sort that the Bible claims to be). To effect such miraculous transformations requires the sort of design that is only known by human beings through the abstract signs and symbols of natural languages. While all other living organisms evidently benefit from the genetic designs upon which their existence depends (or else they would not exist at all), they are not able to become even slightly aware of it. The smartest chimp who ever lived could have no more understanding of the fundamental problems of intelligence than an amoeba. To really enter the fully abstract world on the opposite side of Einstein's Gulf requires a peculiar kind of intelligence - in particular, the kind that the Bible describes as 'the image of God.' It requires the power to gain access to the Logos, i.e., the human language capacity. But to do so requires the special kind of bridge known as intelligence. Now here lies the gauntlet for evolutionary philosophies of all kinds and varieties: Let them show how, if they are able, a concrete object can become, without the intervention of articulate intelligence, an abstract idea.* ▶ [2]

We know that the creation and meaning of information cannot randomly and spontaneously come into existence, because for information to have meaning requires that an intelligence with understanding first existed. Thus it becomes impossible for the creation and meaning of information to occur on its own from a purely naturalistic or materialistic source.

We also know that information is not fully information unless it has meaning. To expand on this, information is only valid when it has a sender or source, and that sender or source intentionally sends the information to a receiver. We call such an action volition, which simply means the ability to make a decision and perform action. Sequentially thinking, the information sent by the sender or source must have meaning in order to be understood by the receiver. Logically, if someone sends information to a receiver, and the receiver has no way to understand or decipher the information (to gain meaning from it), the information proves worthless to the receiver. In summary, in order for information to be valid and have integrity, three things are required:

1. It must have a sender or source.
2. It must have meaning.
3. It must be able to be understood by a receiver.

Without all three of these parts, information is worthless.

Engineer and creationist Werner Gitt, Ph.D., extensively researched and wrote on the topic of information. He states:

◀ *The most important empirical principles relating to the concept of information have been defined in the form of theorems. Here is a brief summary of them:*

1. *No information can exist without a code.*
2. *No code can exist without a free and deliberate convention.*
3. *No information can exist without the five hierarchical levels: statistics, syntax, semantics, pragmatics, and apobetics.*
4. *No information can exist in purely statistical processes.*
5. *No information can exist without a transmitter.*
6. *No information chain can exist without a mental origin.*
7. *No information can exist without an initial mental source; that is, information is, by its nature, a mental and not a material quantity.*
8. *No information can exist without a will.*

We have already shown that life is overwhelmingly loaded with information; it should be clear that a rigorous application of the science of information is devastating to materialistic philosophy in the guise of evolution. ▶ [3]

Information exists in space, in nature, and within living things. Evolution proposes that random chances and unintelligent processes somehow created everything, including the existence of information. But information demonstrates exactly the opposite to what evolution proposes: the sending, receiving, and meaning of information demonstrate definite purpose and intelligence. Information requires an intentional communication, with meaning, from a sender to a receiver, and information involves patterns, codes, and language, all of which are hallmarks of design.

Scientific Reason 59 Summary: Because there are information attributes which could never arise accidently by random chances, and because greater information must first exist in order to produce subsequent information, and because information requires a three-part condition of sender/meaning/receiver, evolution and its premise that information resulted from purely naturalistic and materialistic causes is logically and scientifically impossible.

⌘ Scientific Reason 60 ⌘
Laws of Math

The laws of math are orderly, structured, and consistent. As a very simple example, two times two always equals four. Nobody can reasonably deny or argue against this rule in any logical way. There are many known laws of math, ranging from simple ones such as the rule of subtraction to very complex laws involving advanced formulas. These laws of math are consistent, universal, absolute, and true. They are not dependent on the physical universe behaving in a specific way or on the opinions and preferences of people in order to be true.

Numbers exist as concepts and information; they are not part of the material or physical realm of existence. Furthermore, the laws of math, which of course are dependent upon numbers and the relationships between numbers, are also not part of the material or physical realm of existence. Remember that evolution depends upon a foundational premise that everything which exists has a purely naturalistic or material cause. But this is not true with a number or a law of math, because they are not part of the physical realm of existence. There is no scientific or logical way for something purely physical to create something non-material. Logically, because numbers and the laws of math exist, this foundational premise of evolution is wrong.

Physicist and evolutionist Eugene Paul Wigner, Ph.D., highlights the dilemma faced by evolution in trying to explain the existence of the laws of math:

◀ *The miracle of the appropriateness of the language of mathematics for the formulation of laws of physics is a wonderful gift which we neither understand nor deserve.* ▶ [4]

Scientific Reason 60 Summary: The existence of the laws of math, which are orderly, intelligent, consistent, universal, absolute, and true, is not compatible with evolution and its premise that random chances and purely naturalistic and materialistic processes are somehow the sole explanation for the source of everything, including all laws of math.

⌘ Scientific Reason 61 ⌘
Fibonacci Sequence

Named after Leonardo Pisano Fibonacci, who was born in 1170 A.D. in Pisa, Italy, the Fibonacci Sequence is a pattern of numbers for which adding two sequential numbers gives the next number. The sequence begins as 0, 1, 1, 2, 3, 5, 8, 13, 21, 34, 55, 89, 144, 233, 377, 610, and continues on into infinity. Many remarkable mathematical relationships exist for Fibonacci numbers, such as the ability to sum together any 10 Fibonacci numbers and achieve a resulting number that is divisible by 11. Another relationship begins with the numbers 34 and 55, for which dividing a Fibonacci number by the number which immediately follows it always results in the number 1.618, which has become known as the Golden Ratio.

Examples of Fibonacci Sequence numbers and the Golden Ratio can been seen in the design of plants, animals, and people. In addition to the Golden Ratio, Fibonacci numbers appear as a Golden Spiral, a Golden Angle, and a Golden Rectangle. Because Fibonacci relationships are known to be pleasing in appearance, these patterns and shapes often emerge in art, music, and architecture.

A Golden Spiral is formed when every quarter turn (90 degrees) of an arc moves 1.618 times as far away from the arc's center point as the previous quarter turn. Marvelously, many spirals appearing in nature are Golden Spirals. Golden Spirals can be seen in sunflowers, pinecones, seashells, hurricanes, and spiral galaxies. The Golden Angle occurs as a relationship of 137.5 degrees to 222.5 degrees, which together equal the 360 degrees of a full circle. The ratio of the Golden Angle is a 1 to 1.618 ratio. The Golden Rectangle appears as a rectangle whose ratio of its long side to its short side is 1.618 to 1. Interestingly, if a square, which has sides equal in distance to the width of a Golden Rectangle, is placed within an end of that Golden Rectangle, the remaining rectangle would again be a Golden Rectangle.

We see the Fibonacci Sequence and the Golden Ratio displayed as beautiful patterns in plants and animals:

- Flowers: many flowers display Fibonacci numbers in their petal arrangements. The Trillium has a petal arrangement of 3; the Yellow Violet has a petal arrangement of 5; a Delphinium has a petal arrangement of 8; the Marigold has a petal arrangement of 13; the Helenium has a petal arrangement of 21; the Ox-Eye Daisy has a petal arrangement of 34; the Field Daisy has a petal arrangement of 55; and the Michaelmas has a petal arrangement of 89.

- Pinecones: some pinecones have 5 rows of outer seed scales flowing one direction and 8 rows of outer seed scales flowing in an opposite direction.

- Pineapples: some pineapples have patterns of outer scales arranged in spirals that occur in sets of 5 and 8, and some in sets of 8 and 13.

- Sunflowers: some varieties of sunflowers have spiral patterns of seeds which occur in sets of 8 and 13, some in sets of 21 and 34, some in sets of 34 and 55, some in sets of 55 and 89, and some in sets of 89 and 144.

- Plants: some plants display the Golden Ratio in the spiral pattern displayed by the arrangement of their leaves around the stem. Beginning at one leaf, the number of turns around the stem to arrive at another leaf in the same position equals a Fibonacci Sequence, and the number of leaves between those turns can also display a Golden Ratio. In some elm and basswood trees, the ratio of turns to leaves is 1:2. In some hazel and beech trees, the ratio of turns to leaves is 1:3. In some apricot, oak, and cherry trees, the ratio of turns to leaves is 2:5. In some pear and poplar trees, the ratio of turns to leaves is 3:8. In one almond tree the ratio of turns to leaves is 5:13. Interesting to note, the leaves and stems of some plants are arranged at an angle of 137.5 degrees from each other, reflecting the Golden Angle. Botanists have discovered that this angle maximizes the surface area of the leaves and enables them to receive the greatest amount of sunshine. Even some root systems display the Golden Ratio.

- Chambered Nautilus: the curvature of its shell displays the Golden Spiral.

- Vegetables: a cauliflower, an artichoke, and a variety of broccoli display the Golden Spiral within the shape of the head.

- Honey Bee: the number of bees per generation in a Honey Bee family tree follows the Fibonacci sequence.

- Milky Way Galaxy: some of its spiral arms display the Golden Spiral.

- People: the curvature of our ear rim displays the Golden Spiral. The length of our hand bones display the Golden Ratio, beginning with the finger-tip bone length compared to the length of the next bone section in our finger. For many people, their forearm is about 1.6 times as long as the length of their hand. A DNA molecule displays the Golden Ratio, measuring 34 angstroms by 21 angstroms (an angstrom is equal to 1 hundred-millionth of a centimeter) at its full cycle within the double-helix spiral.

In addition to the many places we see the Fibonacci Sequence and the Golden Ratio displayed as beautiful patterns in living things and our universe, over the years, the Fibonacci Sequence has inspired beauty within the creations of musicians and artists:

- Music: Fibonacci numbers are displayed on a piano. The piano keyboard is designed as an octave of 13 notes, with five black keys and eight white keys. Additionally, the black keys are arranged in sets of two and three. Some symphonies are written based on the Golden Ratio, such that their first movement length is 1.6 times as long as the second movement length.

- Architecture and Art: the Great Pyramid of Giza, the Parthenon of ancient Greece, and the Cathedral of Florence all follow Golden Ratio patterns. Many famous artists have painted with it. For example, Leonardo Da Vinci painted the *Mona Lisa* with her face 1.6 times as long as it is wide, and he painted the dimensions of his *The Last Supper* table, walls, and backgrounds using Golden Ratios.

Scientific Reason 61 Summary: The Fibonacci Sequence appears as Golden Ratios in animals, plants, people, and space. The Fibonacci Sequence displays distinctive order, stunning patterns, and incredible beauty, none of which can be consistently explained by evolution's belief that unintelligent, random, and purely naturalistic processes created order, patterns, and beauty in everything.

⌘ Scientific Reason 62 ⌘ Design, Beauty, and Complexity of DNA

Our DNA reveals intricate design, beauty, order, and complexity that could never be the result of the random, purposeless chance processes promoted by evolution. For example, in just one of the estimated 17 trillion cells within an adult human body, geneticists recognize the amount of information in that one cell's DNA would fill up 1,000 books, each holding 500 pages of normally formatted text. The diameter of a human DNA molecule equals 2 nanometers (2,000 millionths of a meter) and the distance between a pair of nucleotide bases equals .34 nanometers. The length of one strand of human DNA, if it could be uncoiled from its normal double-helix shape, equals nine feet. A DNA molecule is incredibly thin, having a diameter to length ratio equal to 1 to 1,350,000,000. To put this into perspective, if the diameter of a DNA model was 39 inches wide, its total length would be 839,000,000 miles long! [5]

DNA exists as a specific sequence of nucleotide bases. The nucleotide bases in DNA are adenosine, cytosine, guanine, and thymine. Based upon the discovery by geneticist and evolutionist James Watson, Ph.D., and geneticist and evolutionist Francis Crick, Ph.D., we now know that DNA is shaped as a double-helix built with two parallel strands of specifically arranged nucleotide bases. In people, each of these strands are made of three billion nucleotide bases, with each base chemically linked to an opposite base on the parallel strand. A gene is made from specific combinations of nucleotide bases, with some genes made from millions of nucleotide bases.

Consistent with all forms of information, DNA without a language system would be meaningless. In other words, the information coded by the sequence of nucleotide bases is worthless unless it communicates a language and that language is read and understood by something else. One of the fascinating truths about DNA is that the ability for cellular machinery to read and act upon the DNA language (the specific sequence order of the nucleotide bases) is itself encoded by the DNA! Philosopher and evolutionist Sir Karl Popper highlights this amazing truth about DNA:

◀ *What makes the origin of life and of the genetic code a disturbing riddle is this: the genetic code is without any biological function unless it is translated; that is, unless it leads to the synthesis of the proteins whose structure is laid down by the code. But... the machinery by which the cell (at least the non-primitive cell, which is the only one we know) translates the code consists of at least 50 macromolecular components which are themselves coded in the DNA. Thus the code cannot be translated except by using certain products of its translation. This constitutes a baffling circle; a really vicious circle, it seems, for any attempt to form a model or theory of the genesis of the genetic code. Thus we may be faced with the possibility that the origin of life (like the origin of physics) becomes an impenetrable barrier to science, and a residue to all attempts to reduce biology to chemistry and physics.* ▶ [6]

Biochemist and agnostic Michael Denton, Ph.D., performed research in a field of study known as DNA equidistance. This field of study involves measuring the distance between genes on a strand of DNA. A simple analogy here would be to think of a set of rungs on a ladder as a gene, and you could measure and compare the distance between rungs. He discovered that gene spacing distances in bacterial DNA are as different from horses as they are from yeast! The varying distances between the genes identify that DNA is distinct and unique for all creatures, and does not display any evolutionary relationships. If evolution was true, genes would display consistent and uniform patterns of gene spacing in creatures which were more closely evolved to one another. Said another way, if evolution was true, we would observe smaller gene spacing distances between species that are more simplistic and evolutionally related, such as between a single-celled bacteria and a yeast, and we would observe larger gene spacing distances between species that are more distantly evolutionally related, such as between bacteria and horses. [7]

Genetic evolutionists are being inconsistent by holding to a position that DNA supposedly evolved through small, random, chance-based, purposeless, successive genetic processes, but then using the following words in their vocabulary when they discuss elements of DNA: code, protein code, coding regions, encode, decode, codon, anti-codon, transcription, translation, blueprint, program, information, instruction, edit, decipher, messenger, reading, proofreading, signal, alphabet, letter, language, gene expression, and surveillance. None of these words imply any level of randomness or chance. Instead, just the opposite is true - they all of course imply intelligence, purpose, design, and order by the source that originally created the DNA.

Over the last century, genetic evolutionists have proposed that many living things have significant amounts of junk DNA in their genome. This junk DNA was allegedly comprised of genes that now have no purpose but have been left over after years of supposed evolutionary advancements. Today, with many astonishing advancements in genetic research, genetic scientists realize all genes serve a purpose. In fact, in many cases, there are extremely complex interrelationships between multiple genes, such that one gene may play a role in 10 or more different combinations with other genes, and one gene may be acted upon at different times within the life of the creature. So the idea of junk DNA has itself been junked!

The replication of DNA is a truly fascinating process. When a DNA molecule splits apart during the process of DNA replication, it rotates at a speed of 250 revolutions per second, copying 10,000 nucleotide bases each second! A DNA molecule encodes the information needed to build RNA and proteins, but RNA and proteins are required for the replication and translation of the information stored within a DNA molecule. The process of RNA transcription, one of the important steps within DNA replication, is irreducibly complex. All three parts, the fully built strand of DNA, the fully built strand of RNA, and the complete proteins required for successful replication, must all exist and work exactly together in order for the process to be successful. Such a process could never occur by having only DNA, or only RNA, or only proteins evolve first. Geneticist and creationist Jeffrey Tomkins, Ph.D., summarizes this unsolvable challenge for evolution:

◀ *Protein, DNA, and RNA present a complex three-way version of the ancient 'chick or the egg' conundrum. DNA encodes RNA transcripts, but both RNA transcript production and DNA replication are dependent on specific proteins. And proteins are themselves encoded in genes (DNA) and produced from RNA transcripts that are copied from genes in DNA. All three of these major biomolecules need each other in the cell system, so which biomolecule evolved first? This is an irreducibly complex situation that could not have evolved by random naturalistic processes.* ▶ [8]

Biophysicist and evolutionist Caryl Haskins, Ph.D., notes the inability for evolution to explain the process of DNA replication, the translation of replicated DNA, and the activity of highly precise enzymes involved in the replication process:

◀ *But the most sweeping evolutionary questions at the level of biochemical genetics are still unanswered. How the genetic code first appeared and then evolved and, earlier than that, how life itself originated on earth remain for the future to resolve, though dim and narrow pencils of illumination already play over them. The fact that in all organisms living today the processes both of replication of the DNA and of the effective translation of its code require highly precise enzymes and that, at the same time the molecular structures of those same enzymes are precisely specified by the DNA itself, poses a remarkable evolutionary mystery... Did the code and the means of translating it appear simultaneously in evolution? It seems almost incredible that any such coincidence could have occurred, given the extraordinary complexities of both sides and the requirement that they be coordinated accurately for survival. By a pre-Darwinian (or a skeptic of evolution after Darwin) this puzzle would surely have been interpreted as the most powerful sort of evidence for special creation.* ▶ [9]

Scientific Reason 62 Summary: Because of the amount of information stored within one molecule of DNA, and because one double-helix of human DNA is made with six billion specifically sequenced nucleotide bases, and because the building of the cellular machinery needed to read DNA is itself coded by the DNA, and because DNA equidistance relationships display the opposite to what evolution requires, and because the myth of junk DNA has been debunked, and because of the complexity and precision displayed during DNA replication, it is utterly and scientifically impossible for evolution over billions of years to have created the magnificent structure, design, beauty, language, and complexity of DNA.

⌘ Scientific Reason 63 ⌘
DNA Mutations

DNA mutations are strongly promoted as one of the two major pillars upholding evolution. Simply defined, a mutation involves a mistake or alteration in the expected sequence of nucleotide bases within a strand of DNA.

The field of genetics dominates much of the scientific research occurring today. Geneticists observe that genetic mutations from one generation to the next are always neutral or harmful but never create brand new DNA. In fact, genetic scientists now know that mutations are harmful over 99% of the time! A neutral mutation changes the sequence of nucleotide bases but we do not observe any current impact to the gene. A harmful mutation destroys or debilitates one or more genes. It is extremely significant that mutations never build brand new nucleotide sequences (DNA) within the gene pool of any creature, which is required for evolution to be true, but instead, all mutations only destroy or rearrange existing nucleotide base sequences.

Genetic evolutionists will site an example of a beneficial mutation as support for evolution. A beneficial genetic mutation causes a resulting physical change that benefits the creature. A good example is a beetle that normally has wings and lives on a windy island. If the beetle were to experience a mutation that results in the loss of its wings, this would benefit the beetle to some degree, because it would now have less chance of being blown off the island when it flies into a strong wind. But the key here is that the beetle ended up with a net loss of DNA and no more wings, because the mutation caused it to lose the genetic information that produced its wings. So even the example of a beneficial mutation does not scientifically support evolution because the net effect of the mutation produces a creature with less DNA - but a massive gain of brand new DNA is required to support evolution. No genetic research has ever observed a mutation creating new DNA. And to put this point into perspective, we must remember that evolution requires millions of creatures over millions of years each repeatedly, consistently, and successively gaining millions or billions of *brand new* genes!

M.I.T. physicist and creationist Lee Spetner, Ph.D., shares a simple but profoundly realistic reason why mutations cannot be used to support evolution over billions of years:

◀ *Whoever thinks that macroevolution can be made by mutations that lose information is like the merchant who lost a little money on every sale but thought he could make it up on volume.* ▶ [10]

Botanist and evolutionist Michael Pitman, Ph.D., shares that there has never been a scientific experiment in which a DNA mutation produced a brand new gene, a brand new hormone, a brand new enzyme, or a brand new organ:

◀ *Do we, therefore, ever see mutations going about the business of producing new structures for selection to work on? No nascent organ has ever been observed emerging, though their origin in pre-functional form is basic to evolutionary theory. Some should be visible today, occurring in organisms at various stages up to integration of a functional new system, but we don't see them. There is no sign at all of this kind of radical novelty. Neither observation nor controlled experiment has shown natural selection manipulating mutations so as to produce a new gene, hormone, enzyme system or organ.* ▶ [11]

Genetic scientists recognize that mutations are occurring in the DNA of all creatures. Researchers have advanced genetic research technology to the point where they can count single gene differences and then compare these gene differences within different people (such as comparing all the members of an extended family). These single gene differences are known as "Single Nucleotide Variants" or SNVs. Estimates indicate that with the passing of each generation of people, there are 100 new SNVs per generation. When we estimate the number of generations within the past 6,000 years, if each generation length averaged 30 years, this produces 200 generations of people and 20,000 mutations in the human gene pool over the past 6,000 years. There is no exact way currently to understand fully just how many mutations the human genome can endure, but certainly the number 20,000 is quite an impact. Furthermore, geneticists can reasonably deduce that if people had been slowly evolving for the last 2.5 million years, the number of accumulated mutations would be so great that people could not survive today due to the number of harmful SNVs. Here is the math: If people started to evolve about 2.5 million years ago, an average generation length of 30 years equals 83,000 generations and 8,300,000 mutations. This number of mutations in the human genome would not support a normally functioning human body. [12]

In order to better understand the impact of mutations, zoologist and evolutionist Pierre Paul Grasse, Ph.D., studied mutations in hundreds of thousands of bacterial generations. The bacteria in his research multiplied with a generation time of only 30 minutes, which is 400,000 times faster than a human generation. His research, in effect, allowed the study of an equivalent 3.5 million years of human mutational changes. His research showed that bacteria did not change into any other type of creature or develop any brand new DNA during all these generations. It is completely unscientific and illogical to suggest that humans, who are trillions and trillions of times more complex than a single-celled bacteria, have evolved over millions of years when single-celled bacteria have not, especially considering that the first life form according to evolution was a single-celled life form. In his book *Evolution of Living Organisms*, Dr. Grasse states:

◀ *Through use and abuse of hidden postulates, of bold, often ill-founded extrapolations, a pseudoscience has been created. It is taking root in the very heart of biology and is leading astray many biochemists and biologists, who sincerely believe that the accuracy of fundamental concepts has been demonstrated, which is not the case.*

Today, our duty is to destroy the myth of evolution, considered as a simple, understood, and explained phenomenon which keeps rapidly unfolding before us. Biologists must be encouraged to think about the weaknesses of the interpretations and extrapolations that theoreticians put forward or lay down as established truths. The deceit is sometimes unconscious, but not always, since some people, owing to their sectarianism, purposely overlook reality and refuse to acknowledge the inadequacies and the falsity of their beliefs.

It follows that any explanation of the mechanism in creative evolution of the fundamental structural plans is heavily burdened with hypotheses. This should appear as an epigraph to every book on evolution. The lack of direct evidence leads to the formation of pure conjectures as to the genesis of the phyla; we do not even have a basis to determine the extent to which these opinions are correct. ▶ [13]

Theologian and creationist Douglas F. Kelly, Ph.D., identifies that genetic research has observed variation within individual kinds of creatures due to the rearrangement of their existing genes, but one kind of creature developing into another kind of creature because it gained brand new DNA has never been observed:

◀ *Genetic studies indicate that while there is wide variation possible within the broad groups known as 'kinds', there has never yet been demonstrated one kind developing into another, or a more primitive kind developing itself into a higher, more complex form.*

The belief that they do, although widely held by impressive numbers of persons and great institutions for well over a century, is still - scientifically speaking - nothing but that: a belief. It is a fundamental plank in the humanist platform (or paradigm) of materialistic monism, and has to be assumed (just as biogenesis) in order to be proven. ▶ [14]

Geneticist and evolutionist Brian Charlesworth, Ph.D., notes the failure of biology to provide any evidence that mutations have been the cause of living things:

◀ *It should be clear that the claim for an inherent evolutionary increase in entropy and organization is based on an arbitrary model which shows signs of having been constructed simply to yield the desired result... There is nothing in evolutionary or developmental biology that justifies their assumptions that a successful mutation (which seems merely to mean a selectively neutral one in their model) is always associated with an increase in some global measure of phenotype [physical appearance]. Nor is there anything to support the assumption that new species arise as the result of single gene mutations and are initially genetically uniform. If these assumptions are removed, the whole [evolutionary] edifice collapses.* ▶ [15]

Physicist and evolutionist Fritjof Capra, Ph.D., served as the Director of the Center for Ecoliteracy in Berkeley, California. He shares a significant point that mutations do not occur frequently enough to support evolution:

◀ *It has been estimated that those chance errors occur at a rate of about one per several hundred million cells in each generation. This frequency does not seem to be sufficient to explain the evolution of the great diversity of life forms, given the well-known fact that most mutations are harmful and only very few result in useful variations.* ▶ [16]

Scientific Reason 63 Summary: Because DNA mutations are neutral or harmful 99% of the time, and because all mutations never create brand new DNA, and because no scientific experiment has ever observed a mutation creating a brand new gene (or hormone, enzyme, or organ), and because genetic research tells us the human genome could not endure the current amount of observed generational SNVs over millions of years, and because mutations do not mathematically occur frequently enough to support the timeframe required by evolution, DNA mutations scientifically fail to support evolution over billions of years.

Chapter 8 Endnotes

1 – Albert Einstein, *Out of My Later Years* (Secaucus, NJ: Citadel, 1956) pages 59-95.

2 – John W. Oller, Jr., "Einstein's Gulf: Can Evolution Cross it?", *Acts & Facts*, Vol. 9 (2000) page 29. <http://www.icr.org/article/einsteins-gulf-can-evolution-cross-it>

3 – Werner Gitt, "Information, science and biology", *Journal of Creation*, Vol. 10, Iss. 2 (August 1996). <http://creation.com/information-science-and-biology>

4 – Eugene Paul Wigner, "The Unreasonable Effectiveness of Mathematics in the Natural Sciences", *Communications on Pure and Applied Mathematics*, Vol. 13, Iss. 1 (1960) pages 1-14.

5 – Werner Gitt, *Did God Use Evolution* (Green Forest, AR: Master Books, 2006) page 126.

6 – K. R. Popper, "Scientific Reduction and the Essential Incompleteness of All Science", *Studies in the Philosophy of Biology: Reduction and Related Problems* (Berkeley, CA: University of California Press, 1974) page 270.

7 – Nathaniel Jeanson, *Creation Basics and Beyond*, Editor Jason Lisle (Dallas, TX: Institute For Creation Research, 2013) page 140.

8 – Jeffrey Tomkins, *Creation Basics and Beyond*, Editor Jason Lisle (Dallas, TX: Institute For Creation Research, 2013) page 192.

9 – Caryl P. Haskins, "Advances and Challenges in Science in 1970", *American Scientist*, Vol. 59 (May/June 1971) page 305.

10 – Lee Spetner, *Not by Chance! Shattering the Modern Theory of Evolution* (Brooklyn, NY: The Judaica Press, Inc., 1997) page 160.

11 – Michael Pitman, *Adam and Evolution* (London, England: Rider, 1984) pages 67-68.

12 – Brian Thomas, *Creation Basics and Beyond*, Editor Jason Lisle (Dallas, TX: Institute For Creation Research, 2013) pages 141-145.

13 – Pierre-Paul Grasse, *Evolution of Living Organisms* (New York: Academic Press, 1977) pages 6, 8, and 31.

14 – Douglas F. Kelly, *Creation and Change* (Scotland, UK: Christian Focus Publications, 1997) page 174.

15 – Brian Charlesworth, "Entropy: The Great Illusion, a review of Evolution as Entropy by Daniel R. Brooks and E. O. Wiley", *Evolution*, Vol. 40, No. 4 (1986) page 880.

16 – Fritjof Capra, *The Web of Life* (New York: Anchor Books, 1996) page 228.

Chapter 8 Resources

Information

- http://www.icr.org/article/einsteins-gulf-can-evolution-cross-it
- http://creation.com/laws-of-information-1
- http://creation.com/information-science-and-biology
- https://answersingenesis.org/genetics/information-theory/information-science-and-biology
- *The Creation Answers Book* by Creation Ministries International
- *In The Beginning Was Information* by Master Books
- *In Six Days – Why 50 Scientists Choose to Believe in Creation: Chapter 24* by Master Books

Laws of Math

- http://www.icr.org/article/7098
- https://answersingenesis.org/is-god-real/god-natural-law

Fibonacci Sequence

- http://creation.com/golden-numbers
- http://digitalcommons.liberty.edu/honors/334
- http://www.creation-facts.org/tag/fibonacci-numbers

Design, Beauty, and Complexity of DNA

- http://creation.com/genetics-questions-and-answers
- http://www.icr.org/article/junk-dna-myth-continues-its-demise
- http://www.icr.org/article/no-junk-dna
- http://www.icr.org/article/dna-stew-pendous-creation
- https://answersingenesis.org/evidence-for-creation/design-in-nature/dazzling-design-in-miniature
- https://answersingenesis.org/genetics
- *In Six Days – Why 50 Scientists Choose to Believe in Creation: Chapter 18* by Master Books
- *Evolution's Achilles Heel: Chapter 2* by Creation Ministries International
- *Evolution the Grand Experiment: Chapter 17* by Carl Werner

DNA Mutations

- http://creation.com/mutations-questions-and-answers
- http://creation.com/human-genome-decay-and-origin-of-life
- http://creation.com/genetics-and-biblical-demographic-events
- http://www.icr.org/article/158
- https://answersingenesis.org/genetics/mutations/are-mutations-part-of-the-engine-of-evolution
- *The New Answers Book 2: Chapter 7* by Answers in Genesis

⌘ Chapter 9 ⌘

People

And in man is a three-pound brain which, as far as we know, is the most complex and orderly arrangement of matter in the universe. [1]

⌘ Scientific Reason 64 ⌘
Growth Rate of the Human Population

The actual number of people on the earth today contradicts the number of people that should exist if evolution was true.

The human population growth rate in recent times equals about 2% per year. Conservatively, the growth rate throughout human history would be half or less than this percentage, as we know that events such as wars, diseases, and famines have killed large parts of our human population throughout history. Even based on a more conservative human growth rate (half) compared to what is occurring today, starting with just eight people 4,500 years ago, when you do the math this conservative human growth rate leads to an estimated human population size today of seven billion people.

Here is another fascinating way to see this... again, conservatively, if the human population growth rate was only a quarter of the current 2% growth rate, and the human population growth rate has continued for only one million years, when you do the math there would have been over 25,000 generations of people that lived on the earth, based on a generation length of 30-40 years. After this hypothetical one million years, the human population today would be the number 1 followed by over 1,000 zeros. It would be impossible for this many people to all stand on the continents of the earth at the same time! And to understand this another way, in this hypothetical one million years, trillions upon trillions upon trillions of people would have lived and died, and we would see human bones and human graves everywhere we look.

Remember that evolution proposes that human evolution from a supposed early ape-man has been occurring for 2.5 million years. The math in this reason drives home another point why this is statistically and scientifically impossible.

Scientific Reason 64 Summary: Calculating the human population size for a hypothetical one million years, by using a significantly more conservative human population growth rate compared to what we observe today, provides mathematical and scientific support that contradicts evolution's claim that people lived and evolved over the last 2.5 million years.

⌘ Scientific Reason 65 ⌘
Amazing Design, Beauty, and Capabilities in People

While many books describe the amazing abilities of the human body, this reason highlights a few of the remarkable patterns of design, the intricate displays of beauty and order, and the multifaceted capabilities that exist in the human body, none of which could be the result of evolution and its dependency on unintelligent and random chance-based processes forming all life.

The ability for people to speak involves the simultaneous abilities of our vocal cords, our throat cavity and its muscles, our tongue, our brain, and our nervous system, each behaving consistently and in coordination with the principles of air molecule movement and electron motions. Our ability to speak shouts out yet another example of irreducible complexity (See Scientific Reason 55). It is impossible for our ability to speak to be the result of evolution's random, purposeless, and directionless processes, because without all of our speech system parts being fully developed and simultaneously operational, we could never speak. Additionally, the process of speech is a non-material phenomenon. Because evolution rests upon the premise that only material causes can be responsible for forming life, by its own definition, evolution could never produce a non-material phenomenon such as speech and language.

Many biologists and neurologists recognize that the human brain qualifies as the most complicated structure in our universe. A human brain contains over 100 billion nerve cells, with each nerve cell having over 50,000 neurological connections to other brain cells, equaling an estimated 5,000 trillion neurological connections throughout our brain. Our brain receives over 100 million distinctive nerve signals every second from throughout our body. In addition to conscious thought, we can reason, anticipate consequences, and devise plans, all without even consciously knowing about this brain activity. The amazing structure, design, and ability of a human brain is completely incompatible with evolution's beliefs that random and designless processes created something as intricate as the human brain. Biochemist, author, and evolutionist Isaac Asimov, Ph.D., highlights the incredible complexity demonstrated by our brain:

◀ *And in man is a three-pound brain which, as far as we know, is the most complex and orderly arrangement of matter in the universe.* ▶ [1]

Audiologists have estimated that the human inner ear contains as many circuits as found in a metropolitan telecom system. For example, 30,000 hair cells line the cochlea. When a fluid wave within the cochlea causes a vibration, the cochlear nerve cells at the base of each hair send a signal to our brain and our brain interprets these signals so that we "hear". The ear and its thousands of parts have the ability to convert mechanical energy (the sound waves traveling through air into our ear) into hydraulic energy (as the fluid in our cochlea moves) and subsequently convert that hydraulic energy into electrical energy (the electrical nerve signals sent through the vestibulocochlear nerve) that reaches our brain. The process of hearing, of course, could be stated much more thoroughly and medically, but the simple description summarized here confirms that the design and sophistication of our ears certainly could not be the result of the random chance-based processes offered by evolution. Audiologist Gordon Whitehead, M.A., summarizes the beauty and design of our ears:

◀ *The human ear is a rather wondrous instrument. It is composed of tens of thousands of component parts, can work quite flawlessly from well before we are born to more than a century of age, and is capable of performing extremely sophisticated auditory tasks. And, it works 24 hours a day!* ▶ [2]

Our eyes shine light on more demonstrations of beauty and complexity in the human body. There are over 100 million light-sensitive nerve endings in a human eye that constantly send electrical signals to our brain, and these signals travel through one million nerve fibers located in each of our two optic nerves. Ophthalmologists estimate that a human eye sends eight million electrical impulses to the brain every second. In just one day, a human eye moves 100,000 times. Most interestingly, even Charles Darwin himself highlights the absurdity that our eyes could be the result of evolution:

◀ *To suppose that the eye with all its inimitable contrivances for adjusting the focus to different distances, for admitting different amounts of light, and for the correction of spherical and chromatic aberration, could have been formed by natural selection, seems, I freely confess, absurd in the highest degree.* ▶ [3]

A human heart can beat 400 million times in a year. In an average lifetime, a human heart will pump over 500,000 metric tons of blood through its circulatory system. In an average person, there are over 100,000 miles of veins, arteries, and capillaries, enough to circle the earth's equator about four times! [4] The impressive structure, design, and ability of a human heart and circulatory system is not compatible with the random chance-based processes offered by evolution.

There are hundreds of different perfectly regulated chemical reactions involved within the metabolism of a human body. The simultaneous interactions and precise coordination of these chemical reactions, which all must interact and coordinate in specific sequences in order to properly function, could never be the result of the random-chance processes offered by evolution.

Our immune system involves the synchronized coordination of skin, bone marrow, thymus, lymph nodes and fluids, spleen, blood and white blood cells, antibodies, and hormones. Working together in a highly detailed and orderly fashion, the parts of our immune system constantly work together to keep us healthy. For example, biologists estimate we each have over 50 billion white blood cells in our bloodstream protecting us against unwanted bacteria and disease. Evolution cannot explain the irreducible complexity built into our immune system.

The human reproductive system also provides a powerful example of irreducible complexity. It would be impossible for our reproductive systems to be the result of evolution because two fully developed and functioning reproductive systems (male and female) are the prerequisite to produce another person with a fully functioning reproductive system. Said another way, it is impossible for our reproductive systems to work unless both the male and female reproductive systems are functioning and existing simultaneously. And related to this point, consider the fact that every person begins merely as a single cell. After conception, this single cell follows a process of specialized and precise cell division that eventually leads to a full-grown body made of trillions of cells each having a specific purpose, location, and function. It is just not scientifically reasonable or statistically possible that the precise and specialized cell division processes transforming one cell into a body with trillions of specialized cells could be the result of random chance!

The blood clotting system in our bodies involves a complicated set of chemicals and processes to keep us from bleeding to death when our skin is cut. It forms an irreducibly complex system that can only work if all of the parts in the system are available and working correctly. A blood clot builds from a collection of platelets that entangle themselves into a network of fibrin molecules within the constricted blood vessel at the location of the cut or damage. Blood clotting requires the enzyme thrombin, specific proteins, calcium ions, and about a dozen other clotting factors that are freely circulating within our blood stream. Medical doctor and creationist Andrew Hodge, M.D., sums up the significance of how our blood clotting system functions, highlighting it could not be the product of small, continual, successive, genetic and physical changes but instead demonstrates irreducible complexity:

◀ *The function of the blood clotting system is to prevent the escape of blood from a damaged vessel. To do this, the blood has a special and very complex repair procedure in place. Once initiated by a cut, the first component in the process is activated, which in turn activates the next component, and so on, in a series of cumulative, mutually-dependent steps. This physiological chain of production, or cascade, results in the formation of a solid obstruction (a clot) in order to seal over the damage.*

Some of the main components of the clotting cascade are the proteins fibrinogen, prothrombin, Stuart (anti-hemophilic) factor and proaccelerin. None of these are used for any other purpose in the blood. The system is very finely tuned to result in a repair process that achieves just the repair needed at just the right place and time to stop bleeding and begin the process of healing. Importantly, the process is also self-limiting to ensure that coagulation (clotting) of the entire blood supply does not occur. ▶ [4]

And finally, as an example of the fantastic demonstration of beauty, design, and functionality in our body, neurosurgeon Ben Carson, M.D., describes the incredible and complex activities that occur within our body when someone is asked to perform the simple task of raising his hand:

◀ *Let's think about what your brain had to do when I asked you that question. First of all the sound waves had to leave my lips, travel through the air into your external auditory meatus, travel down to your tympanic membrane, set up a vibratory force which traveled across the ossicles of your inner ear to the oval round windows, generating a vibratory force in the endolymph, which mechanically distorted the microcilia, converting mechanical energy into electrical energy, which traveled across the cochlear nerve to the cochlear nucleus at the ponto-medullary junction, from there to the superior olivary nucleus... ascending bilaterally to the brain stem through the lateral lemniscus to the inferior colliculus and the medial jeniculate nucleus, across the thalmic radiations to the posterior temporal lobes to begin the auditory process, from there to the frontal lobes, coming down the tract of Vicq d' Azur, retrieving the memory from the medial hippocampal structures and the mammillary bodies, back to the frontal lobes to start the motor response at the Betz cell level, coming down the cortico-spinal tract, across the internal capsule in the cerebral pednucle, descending down to the cervico-medullary decussation into the spinal cord gray matter, synapsing, going out to the neuromuscular junction, stimulating the nerve and the muscle so you could raise your hand.* ▶ [5]

Scientific Reason 65 Summary: The incredible examples of order, design, beauty, and interrelated functionality demonstrated within the human body are impossible to explain by the mere random-chance, directionless, purposeless processes offered by evolution. It is impossible to explain the existence of irreducibly complex body systems involving our speech capability, our brain and nervous system, our ears, our eyes, our heart and circulatory system, our metabolism, our immune system, our reproductive system, and our blood clotting ability as results of small, successive, continual, genetic and physical changes over millions of years, as required if evolution was true.

⌘ Scientific Reason 66 ⌘
The Energy Currency of Life

Adenosine Triphosphate (ATP) is an extremely small but crucial molecule, made of thousands of atoms, that provides the energy needed by our cells. It has been coined the "energy currency of life", for without it, life would not exist.

The enzyme ATP Synthase builds molecules of ATP with the use of its tiny rotary motor. This microscopic motor can be found embedded within the membrane of a cellular organelle called the Mitochondria, and it produces three ATP molecules for every 10 protons that zip through its center. Microbiologists estimate that an ATP Synthase rotary motor spins at a rate of 9,000 RPMs! [6]

The ATP Synthase enzyme consists of a complex arrangement of proteins, constructed of at least 29 separately made synchronized protein subunits. Incredibly, in the process of providing energy that is burned by almost every one of our body's 17 trillion cells, the chemical reactions performed by the ATP Synthase enzyme to build ATP molecules occur about a million times each minute! All the combined ATP Synthase enzymes in a human body produce an amount of ATP each day that is equal to half of a person's own body weight [6]

Scientific Reason 66 Summary: The incredible design, complexity, order and beauty demonstrated by the ATP Synthase enzyme to produce massive amounts of ATP molecules provides an irreducibly complex system that could never be the result of small, successive, intermediate genetic and physical steps offered by evolution over billions of years.

⌘ Scientific Reason 67 ⌘ Vestigial Organs

Some evolutionary scientists have proposed that certain organs in our bodies are vestigial, meaning these organs no longer serve any purpose and are merely left-overs from the hypothetical evolutionary advancements within our bodies. This is certainly not a scientifically logical proposal. Why? Having vestigial organs that no longer function demonstrates a reduction in the amount of operational DNA which codes for working body organs. In order to be scientifically true, evolution would require a continual massive increase in brand new operational DNA that codes for the newly developing body parts and their functions. The proposal of vestigial organs, which actually represents a reduction of operational DNA in our human genome, implies specifically the opposite to this evolutionary requirement.

The idea of vestigial organs logically implies there are fewer functioning organs in our bodies today than there were a hypothetical million years ago. Therefore, if these organs were really vestigial, this implies "de-evolution", so to speak, because this means operational DNA and its ability to build working organs that actively function is leaving our gene pool.

There have been over 180 supposed vestigial organs proposed over the last couple centuries, and today, the medical community has discovered one or more specific purposes for each of these organs previously considered unneeded. Some commonly known examples include our appendix, our tonsils, our tail bone, our wisdom teeth, and our thymus. We all need to be careful and not allow our ignorance regarding the purpose of a specific body part to lead us to believe that the part has no purpose. [7]

Scientific Reason 67 Summary: The proposal that people have vestigial organs implies that the human genome is losing operational DNA rather than gaining brand new DNA, which is actually the opposite of what evolution requires to be true. Recent scientific research has discovered at least one specific purpose within our bodies for every alleged vestigial organ, giving us another observable and scientific reason why evolution over billions of years is not true.

✍ Chapter 9 Endnotes ✍

1 – Isaac Asimov, "In the Game of Energy and Thermodynamics You Can't Break Even", *Smithsonian Institute Journal* (June 1970) page 10.

2 – Gordon Whitehead, "A Brief Journey Through the Ear", *Digital Recordings* (2010). <http://www.digital-recordings.com/audiocd/journey.html#eighth>

3 – Charles Darwin, *On the Origin of Species by Means of Natural Selection, or the Preservation of Favoured Races in the Struggle for Life* (London, England: J.M. Dent & Sons Limited, 1971) page 167.

4 – Andrew Hodge, "Life is in the blood", *Creation Magazine*, Vol. 33, Iss. 3 (July 2011). <http://creation.com/life-is-in-the-blood>

5 – Bill Barrett, "The National Prayer Breakfast", *Capitol Words*, Sunlight Foundation, Pub 30April1997. Web 13September2015. <http://capitolwords.org/date/1997/04/30/H2054_the-national-prayer-breakfast>

6 – Brian Thomas, "ATP synthase: majestic molecular machine made by a mastermind", *Creation Magazine*, Vol. 31, Iss. 4 (September 2009) pages 21-23. <http://creation.com/atp-synthase>

7 – Randy J. Guliuzza, *Creation Basics and Beyond*, Editor Jason Lisle (Dallas, TX: Institute For Creation Research, 2013) pages 111-117.

Chapter 9 Resources

Growth Rate of the Human Population

- http://creation.com/human-population-growth
- http://creation.com/where-are-all-the-people
- http://www.icr.org/article/recent-explosion-human-diversity

Amazing Design, Beauty, and Capabilities in People

- *Creation Basics and Beyond: Chapter 14* by Institute For Creation Research
- http://www.icr.org/article/5784/360
- http://www.icr.org/article/all-or-nothing-unity
- http://www.icr.org/article/4623/293
- http://creation.com/the-human-body-gods-masterpiece
- http://creation.com/excellent-eye-better-than-any-camera-the-eyes-response-to-light
- http://creation.com/life-is-in-the-blood
- https://answersingenesis.org/human-body/our-impressive-immune-system-more-than-a-defense
- http://www.digital-recordings.com/audiocd/journey.html#eighth
- http://capitolwords.org/date/1997/04/30/H2054_the-national-prayer-breakfast

ATP - The Energy Currency of Life

- http://creation.com/atp-synthase
- http://creation.com/design-in-living-organisms-motors-atp-synthase
- http://www.icr.org/article/sophisticated-protein-motor-defies

Vestigial Organs

- http://creation.com/do-any-vestigial-organs-exist-in-humans
- http://creation.com/vestigial-arguments-remnants-of-evolution
- https://answersingenesis.org/human-body/vestigial-organs

⌘ Chapter 10 ⌘

Evolution

I grew up believing in this (evolution) myth and I have felt - I still feel - it's almost perfect grandeur. Let no one say we are an unimaginative age: neither the Greeks nor the Norsemen ever invented a better story. But the myth asks me to believe that reason is simply the unforeseen and unintended by-product of a mindless process at one stage of its endless and aimless beginning. The content of the myth thus knocks from under me the only ground on which I could possibly believe the myth to be true. If my own mind is a product of the irrational, how shall I trust my mind when it tells me about evolution? [1]

⌘ Scientific Reason 68 ⌘ Existence of Non-Material Things

It is not worthwhile to question whether non-material things exist, as nobody can reasonably deny the reality of morals, human conscience, reasoning, comprehension, ethics, a sense of right and wrong, art, music, drama, and humor. It is impossible for non-material things to have come into existence from purely material causes (those based on matter only) because there is no way for a purely material cause to create anything non-material.

Evolution holds that everything, including non-material things, originated by naturalistic random-chance processes working purely with matter. Contrary to this belief, our universe follows logical and orderly laws, and logic and order cannot be produced by random chance. Even the ability to understand these words, or anything else for that matter, is only possible by using the principles of reasoning and logic.

Famous theologian C. S. Lewis makes two extremely dissecting points about the inability of purely material things to create non-material things:

◀ *If the solar system was brought about by an accidental collision, then the appearance of organic life on this planet was also an accident, and the whole evolution of man was an accident too. If so, then all our present thoughts are mere accidents - the accidental by-product of the movement of atoms. And this holds for the thoughts of the materialists and astronomers as well as for anyone else's. But if their thoughts - i.e. of materialism and astronomy - are merely accidental by-products, why should we believe them to be true? I see no reason for believing that one accident should be able to give me a correct account of all the other accidents.*

I grew up believing in this (evolution) myth and I have felt - I still feel - it's almost perfect grandeur. Let no one say we are an unimaginative age: neither the Greeks nor the Norsemen ever invented a better story. But the Myth asks me to believe that reason is simply the unforeseen and unintended by-product of a mindless process at one stage of its endless and aimless beginning. The content of the myth thus knocks from under me the only ground on which I could possibly believe the myth to be true. If my own mind is a product of the irrational, how shall I trust my mind when it tells me about evolution? ▶ [1]

Scientific Reason 68 Summary: No process, scientific or otherwise, exists for anything non-material to come into existence purely from a material cause, and thus, evolution cannot scientifically explain the origin and existence of everything that is non-material.

⌘ Scientific Reason 69 ⌘
Naturalistic Assumption

Evolution and the idea of a billion-year-old universe flows from a naturalistic assumption, which originates from the belief that only purely material processes (those involving only matter) are responsible for the origin and existence of everything. A simple definition of naturalism is a belief system which only considers physical events or objects (matter) and refuses to recognize any supernatural influence.

Immunologist Scott C. Todd, Ph.D., highlights the thinking that some scientists hold about this assumption:

◀ *Even if all the data point to an intelligent designer, such a hypothesis is excluded from science because it is not naturalistic.* ▶ [2]

Holding a naturalistic assumption about the origin of everything in our universe is not truly scientific because there is no scientific method which can accurately test, observe, or verify this assumption. Additionally, there is nothing inherent or required in the field of science that says only naturalistic reasons must be considered and supernatural reasons cannot be considered. Such a thinking restriction is the result of a personal bias imposed by only some people. Because a personal bias by definition is not scientific, allowing any personal bias to have control over what is observed and acknowledged is in itself not practicing real science. We all should remember that the field of science simply involves the search for knowledge, and restricting how and where we search for that knowledge is not part of the real definition of science.

It is significant to remember that the vast majority of scientific discoveries and advancements occurring before 1900 were made by men and women who believed in the supernatural. One of the greatest scientists of history, Sir Isaac Newton, states:

◀ *This most beautiful system of the sun, planets and comets, could only proceed from the counsel and dominion of an intelligent and powerful Being.* ▶ [3]

Scientific Reason 69 Summary: The idea that the age of our universe is billions of years old flows from a naturalistic assumption which is based upon an unscientific personal bias and is not scientifically testable or supportable.

⌘ Scientific Reason 70 ⌘
Statistical Impossibility

It is statistically impossible for evolution to have occurred.

Mathematicians recognize that 1 chance in $10^{(50)}$, written as 1 chance in 100,000,000,000,000,000,000,000,000,000,000,000,000,000,000,000,000,000, is so near impossible that it has become known as the threshold of possibility. Any event with odds greater than 1 in $10^{(50)}$ is recognized to be impossible, regardless of the amount of time or number of opportunities the event has to occur. Astronomer and mathematician Sir Fred Hoyle, Ph.D., was an honorary research professor at Manchester University and a math lecturer at Cambridge University. He identifies why evolution is statistically impossible:

◀ *I don't know how long it is going to be before astronomers generally recognize that the combinatorial arrangement of not even one among the many thousands of biopolymers on which life depends could have been arrived at by natural processes here on the earth. Astronomers will have a little difficulty at understanding this because they will be assured by biologists that it is not so, the biologists having been assured in their turn by others that it is not so. The 'others' are a group of persons who believe, quite openly, in mathematical miracles. They advocate the belief that tucked away in nature, outside of normal physics, there is a law which performs miracles (provided the miracles are in the aid of biology). This curious situation sits oddly on a profession that for long has been dedicated to coming up with logical explanations of biblical miracles... It is quite otherwise, however, with the modern mathematical miracle workers, who are always to be found living in the twilight fringes of thermodynamics.*

At all events, anyone with even a nodding acquaintance with the Rubik cube will concede the near-impossibility of a solution being obtained by a blind person moving the cubic faces at random. (There are 4 x $10^{(19)}$ possible scramblings of the Rubik cube, that is, ten billion billion). Now imagine $10^{(50)}$ blind persons each with a scrambled Rubik cube, and try to conceive of the chance of them all simultaneously arriving at the solved form. You then have the chance of arriving by random shuffling of just one of the many biopolymers on which life depends. The notion that not only the biopolymers but the operating program of a living cell could be arrived at by chance in a primordial organic soup here on the Earth is evidently nonsense of a high order. Life must plainly be a cosmic phenomenon. ▶ [4]

Biologist and evolutionist Edwin Conklin shares a candid perspective about this statistical impossibility:

◀ *The probability of life originating from accident is comparable to the probability of the unabridged dictionary resulting from an explosion in a printing shop.* ▶ [5]

Mathematician, theologian, and philosopher William Dembski, Ph.D., identifies another way that evolution is not scientifically or mathematically supportable. In an email titled "Who's Got the Magic?" in which he references Phillip Johnson's points on intelligent design and Robert Pennock's book *Tower of Babel*, Dr. Dembski states that evolution is built upon an unscientific type of magical thinking that promotes the belief that everything came from nothing:

◀ *This form of magic... is the view that something can be gotten from nothing... the transformation of nothing into something may involve minor expenditures of effort. For instance, the magician may need to utter 'abracadabra' or 'hocus pocus'. The Darwinian just-so stories that attempt to account for complex, information-rich biological structures are likewise incantations that give the illusion of solving a problem but in fact merely cloak ignorance... We have become so accustomed to this something for nothing way of thinking that we no longer appreciate just how deeply magical it is.* ▶ [6]

In his book *Darwin was Wrong: A Study in Probabilities,* mathematician and non-evolutionist I. L. Cohen shares the mathematical improbability that evolution could even occur:

◀ *To propose and argue that mutations even in tandem with 'natural selection' are the root-causes for 6,000,000 viable, enormously complex species, is to mock logic, deny the weight of evidence, and reject the fundamentals of mathematical probability.* ▶ [7]

Scientific Reason 70 Summary: It is statistically impossible for evolution over billions of years to have occurred.

⌘ Scientific Reason 71 ⌘
Natural Selection

While many evolutionary scientists promote the idea that natural selection provides one of the primary pillars supporting evolution, natural selection actually fails to support evolution. Natural selection can allow a creature to survive better in a particular environment, but it can only decrease the amount of existing DNA or rearrange the existing genes within a creature. Natural selection cannot be the originator of brand new DNA, which would be required for a single cell to morph (evolve) into a fully developed plant, animal or human.

When describing natural selection, some scientists will say that it "selects" or "operates on" or "chooses" or "favors" or "provides" certain advantages, but all of these actions imply the result of intelligent decision making. They are not actions which result from random, unguided, purposeless chance factors, which in reality are the only mechanisms available from evolution. Even the term "selection" is itself a misnomer because natural selection cannot "select" anything because it has no volition or will.

Many biological systems are made of multiple independent parts that combine together to form a very complex system. For example, human muscles, tendons, ligaments, bones, nerves, and circulatory system all work together and fulfill individual roles that help us walk or run. Natural selection is touted as a primary cause of evolution, supposedly weeding out the unfit or useless parts in living things while "selecting" the best and most functional parts. A problem in logic thus surfaces: if natural selection worked as proposed, the non-functioning and undeveloped individual body parts would be eliminated before they could ever become part of a fully complex and functioning system. Until the individual parts are fully developed, fully functioning, and actually contributing to a complex system, they do not work or serve any purpose, and thus, natural selection by its own definition would remove these unneeded or nonworking body parts before they ever had a chance to evolve.

Paleontologist and evolutionist Stephen Jay Gould, Ph.D., comments on the inability of natural selection to account for the intermediate stages of supposedly evolving structures:

◀ *We can readily understand how complex and full developed structures work and owe their maintenance and preservation to natural selection - a wing, an eye, the resemblance of a bittern to a branch or of an insect to a stick or dead leaf. But how do you get from nothing to such an elaborate something if evolution must proceed through a long sequence of intermediate stages, each favored by natural selection? You can't fly with 2% of a wing or gain much protection from an iota's similarity with a potentially concealing piece of vegetation. How, in other words, can natural selection explain these incipient stages of structures that can only be used (as we now observe them) in much more elaborated form?* ▶ [8]

Scientific Reason 71 Summary: Natural selection, by its own definition, cannot be the originator of brand new DNA within a creature and thus, it cannot cause a single cell to evolve into a fully developed plant, animal, or human. Additionally, by its own definition, natural selection would eliminate any intermediate, nonworking structures before they became fully functional. Therefore natural selection scientifically fails to support evolution and its requirement for the development of complex life forms from "simpler" forms.

⌘ Scientific Reason 72 ⌘ Speciation

Evolution proposes that speciation, the process in which a new species arises from another already-existing species, is proof for evolution. But with the modern advancements in genetic research and our recent ability to build genome maps, geneticists have discovered that in every instance of observed speciation, the newly formed species actually has less DNA than its parents did!

In order for evolution to be true, speciation would have to create billions of brand new genes accompanied by an increase in the types of those genes (what the genes code for), but this is never scientifically observed. Geneticist and creationist Georgia Purdom, Ph.D., summarizes why speciation cannot be used to support evolution:

◀ *Speciation has never been observed to form an organism of a different kind, such as a dog species producing a cat. Speciation works only within a kind. Evolution requires natural selection and speciation to give rise to new kinds from a former kind (e.g., dinosaurs evolving into birds). Speciation, however, leads to a loss of information, not the gain of information required by evolution. Thus, speciation as a possible outcome of natural selection cannot be used as a mechanism for molecules-to-man evolution.* ▶ [9]

Scientific Reason 72 Summary: Speciation never creates a new species that has brand new genes, but rather it always results in a reduction in number or the reshuffling of existing genes within a new species. Evolution requires the creation of billions of brand new genes to "evolve" a new creature from an original single cell, but speciation scientifically and genetically fails to offer this capability to evolution.

⌘ Scientific Reason 73 ⌘ Haeckel's Biogenetic Law

In the 1860s, zoologist and evolutionist Ernst Haeckel published pictures claiming to show various creatures during their embryonic development. He offered that his pictures showed a human embryo progressing through its past evolutionary stages during its development within the womb. Many evolutionists accepted his claims, and his idea became coined as Haeckel's Biogenetic Law.

Ernst Haeckel stated that during our embryonic development we display a yolk sac, from our days of being chickens; tails, from our time as a lizard; and gill slits, from our time as a fish. Biologists now know that we never have a yolk sac like a chicken; rather, we have an organ called a "blood forming sac" which is instrumental in the development of our first red blood cells. We of course never have tails, and now we know the small fold of skin erroneously called a "tail" actually plays a key role in the development of muscles and our coccyx bone. And we now know we do not ever have gill slits or any DNA which codes for gills slits; rather, we have a series of throat pouches needed for the development of our thymus gland and our middle ear canals.

Ernst Haeckel's drawings have been shown by numerous scientists to be fraudulent, and our current understanding of human embryonic development has also shown his supposed law to be false. Yet his drawings and their false information continue to be presented as fact in middle-school, high school, and college science books. This deception is certainly not based upon scientific observation, and even Ernst Haeckel himself actually admitted later in his life that he had faked his pictures. Biologist and evolutionist Walter Bock, Ph.D., notes the reality of these fake pictures:

◀ *Moreover, the [Haeckel's] biogenetic law has become so deeply rooted in biological thought that it cannot be weeded out in spite of its having been demonstrated to be wrong by numerous subsequent scholars.* ▶ [10]

Sociologist and evolutionist Jeremy Rifkin highlights that there is not a single biologist in the world that would support Haeckel's Biogenetic Law:

◀ *Haeckel's argument, which is sometimes referred to as the 'biogenetic law,' is a myth. There is not a single prominent biologist in the world who is willing to extend so much as a scintilla of credence to it.* ▶ [11]

Evolutionary scientists will often refer to Haeckel's Biogenetic Law using the expression "ontogeny recapitulates phylogeny", which basically offers scientific wording to claim that a human embryo passes through its supposed evolutionary stages during its time in the womb. Paleontologist and evolutionist George Gaylord Simpson, Ph.D., and medical doctor and evolutionist William Beck, M.D., in their textbook *An Introduction to Biology,* provide a candid concession:

◀ *Haeckel misstated the evolutionary principle involved. It is now firmly established that ontogeny does not repeat phylogeny.* ▶ [12]

Biologist and evolutionist Keith S. Thompson, Ph.D., emphasizes that Haeckel's law has been known to be a fraud for almost 100 years:

◀ *Surely the biogenetic law is as dead as a doornail. It was finally exercised from biology textbooks in the [19] fifties. As a topic of serious theoretical inquiry, it was extinct in the [19] twenties.* ▶ [13]

Scientific Reason 73 Summary: Haeckel's biogenetic law has been genetically, scientifically, and observationally shown to be a fraud, and therefore cannot be used to support belief in evolution.

⌘ Scientific Reason 74 ⌘
Extinction of Dinosaurs

The always-popular reference to the extinction of dinosaurs comes up as an example that allegedly supports evolution. Many people may not realize that the word dinosaur is a relatively new word which was first used by paleontologist Sir Richard Owen in 1841. Prior to this time, such animals were commonly called dragons or behemoths or serpents.

Animals of all kinds, including those recently labeled as a "dinosaur", become extinct. Why is that? The specific causes vary with each animal, but usually include one or more of these reasons: overhunting, lack of food and habitat, diseases, genetic mutations, severe storms, and natural disasters. But when asked specifically why some dinosaurs have gone extinct, many evolutionists will say we cannot know for sure, but yet they are sure dinosaurs went extinct 70 million years ago. Well, the real answer is that they have gone extinct for all the same reasons as other animals that have gone extinct, and there is no mystery.

Paleontologist Mary Schweitzer, Ph.D., and her research team discovered still-soft red blood cells and blood vessels within Tyrannosaurus Rex bones, which were supposedly 65 million years old. Scientifically, it is chemically impossible for red blood cells and vessels to survive for 65 million years, regardless of whether they were fossilized or frozen or influenced by any other decay-reducing factors. During an interview about her discovery, Dr. Schweitzer shares that many in the scientific community refuse to recognize her data and to believe her discovery is true:

◀ *'I had one reviewer tell me that he didn't care what the data said, he knew that what I was finding wasn't possible', says Schweitzer. I wrote back and said, 'Well, what data would convince you?' And he said, 'None.'* ▶ [14]

In addition to Dr. Schweitzer's spectacular discovery, other scientists have found still-soft tissues within dinosaur fossils, including osteocalcin, DNA, hemoglobin, skin, keratin, collagen, retinal tissue, and cartilage. See a link in the chapter resources for this reason for a site listing many of these incredible discoveries during the last few decades.

Scientific Reason 74 Summary: The extinction of dinosaurs is a popular idea used to support evolution over millions of years. But scientifically, because of the recent discovery of many types of still-soft tissues within dinosaur fossils and the known chemical and molecular limitations for tissue decay resistance, we now have scientific research confirming that dinosaurs lived thousands of years ago and not millions of years ago.

⌘ Scientific Reason 75 ⌘
Homologous Structures

One of evolution's claims is that the degree of similarity between the same body structures in different creatures indicates the degree of evolutionary relationship between those creatures. In summary, the claim follows this thinking: if the same body parts (such as an arm or leg) of different creatures appear similar in size and shape, then those creatures must be closely related to a common evolutionary ancestor. But if the same body parts do not appear very similar, then the creatures are more distantly related to a common evolutionary ancestor which lived many more years back in time. Such thinking is a huge assumption which cannot be factually or scientifically substantiated.

With the increases in embryonic research and its associated discoveries, scientists have discovered that these alleged homologous body structures actually develop differently as embryos, which is just the opposite to what should be observed if these structures were the result of evolution. Embryologist and evolutionist Sir Gavin de Beer wrote a paper in 1971 called *Homology: an Unsolved Problem* in which he listed many examples of supposed homologous structures that actually develop differently as embryos. He was never able to resolve this problem before he died, and neither have any other evolutionists since his book was released. [15]

Geneticist and evolutionist Gunter Wagner, Ph.D., comments on the unresolvable reality that observed embryology research contradicts the idea that homologous structures display close evolutionary links:

◀ *The disturbingly many and deep problems associated with any attempt to identify the biological basis of homology have been presented repeatedly.* ▶ [16]

Medical doctor and creationist Jerry Bergman, M.D., summarizes other reasons why the argument of homology has failed to support evolution from a biochemical and molecular biological perspective:

◀ *The recent information explosion in embryology, microbiology, genetics and especially molecular biology has revealed in minute detail how plants and animals are constructed at the molecular level. If the Darwinian interpretation of homology were correct, then we would expect that the same homologies found at the macroscopic level also exist at the microscopic, biochemical and genetic levels. What researchers in each of these fields often find has greatly undermined the homology concept. So many exceptions now exist that molecular biologist Michael Denton concluded that the homology theory should be rejected. His main argument is that genetic research has not shown that homologous structures are produced by homologous genes and follow homologous patterns of embryological development. Instead, genetics has found that homologous structures are 'often specified by non-homologous genetic systems' and furthermore, the homology 'can seldom be extended back into embryology.'* ▶ [17]

Scientific Reason 75 Summary: The idea that homologous structures in living creatures support evolution cannot be conclusively or scientifically substantiated, and recent scientific advancements in embryology and genetic research have confirmed that assumed homologous structures fail to show the required genetic relationships for evolution to be true.

⌘ Scientific Reason 76 ⌘ Morphological Tree Diagrams

Diagrams named "morphological trees" are commonly used to support evolution, but in reality, these diagrams fail to correctly map actual scientific observations and fossil discoveries.

Morphology refers to physical characteristics and appearances, such as bone arrangement, arm length, or eye shape. Morphological tree diagrams supposedly map the evolutionary relationships between living creatures based upon measurements and appearances of their body parts. Evolutionary scientists thought this would be strong evidence for evolution, but just the opposite has actually come to light as scientific research has accumulated.

Contrary to the expectations of evolutionary beliefs, many creatures have body parts similar in structure, size, and shape to the same body parts on other creatures, while at the same time having other body parts that are completely different in structure, size, and shape compared to those on the same other creature. The eye ball of many mammals compared to the eye ball of an octopus offers a significant example. The eyes in both mammals and octopi have similar structure, function, and shape, but obviously for most mammals, not too many other body parts are similar to an octopus!

The belief in morphological tree diagrams grew into the idea that an evolutionary "tree of life" could be diagrammed. But today, biologist and evolutionist Michael R. Rose, Ph.D., acknowledges that an evolutionary tree of life is no longer scientifically recognized:

◀ *The tree of life is being politely buried - we all know that.* ▶ [18]

Physicist and evolutionist Jerald Loewenstein, Ph.D., and anthropologist and evolutionist Adrienne Zihlman, Ph.D., explain that not one single morphological tree diagram exists for which all evolutionary scientists agree, which strikes at the scientific credibility that such a morphological tree even exists:

◀ *But anatomy and the fossil record cannot be relied on for defining evolutionary lineages. Yet paleontologists persist in doing just this. They rally under the banner of a methodology called cladistics, in which family trees of living and fossil primates are constructed on the basis of 'primitive' and 'derived' traits (mostly of teeth and bones), which are either shared or not shared. Shared primitive characteristics are shared because they come from a common ancestor, unshared derived characteristics reveal separate evolutionary paths. The subjective element in this approach to building evolutionary trees, which many paleontologists advocate with almost religious fervour, is demonstrated by the outcome: there is no single family tree on which they agree. On the contrary, almost every conceivable combination and permutation of living and extinct hominoids has been proposed by one cladist or another.* ▶ [19]

Just as significant, all the required parts of a morphological evolutionary tree do not exist and have never been found, even after years of scientific research. Paleontologist and evolutionist Colin Patterson, Ph.D., shares the scientific reality that a morphological tree diagram could never be accurately drawn:

◀ *We have access to the tips of a tree; the tree itself is theory and people who pretend to know about the tree and to describe what went on with it, how the branches came off and the twigs came off, are I think, telling stories.* ▶ [20]

Famous paleontologist and evolutionist Mary Leakey, Ph.D., and her husband Louis Leakey, were noted for making fossil discoveries in Africa. In an Associated Press Interview on December 10th, 1996, Dr. Mary Leakey speaks about the lack of any conclusive scientific morphological tree diagrams:

◀ *All those trees of life with their branches of our ancestors, that's a lot of nonsense.* ▶ [21]

Scientific Reason 76 Summary: Scientific research has not provided the required observations to enable the drawing of a scientifically accurate evolutionary morphological tree diagram, which directly contradicts evolution over long ages.

⌘ Scientific Reason 77 ⌘
No Conclusive Evidence

Although many scientists promote evolution as a fact, no conclusive evidence for evolution exists. This final scientific reason includes significant examples why evolution fails to be a scientific fact.

Identifying a scientific fact using real science involves making a series of observations, forming a hypothesis about what is observed, performing multiple tests to see if that hypothesis is always correct in every experiment, and then recognizing that the hypothesis is always true and can be considered a scientific fact. Directly contrary to this scientific method process, evolution has actually never been observed. Atheist and evolutionist Richard Dawkins provided an eye-opening quote during a TV interview with Bill Moyers on the PBS network in December 2004. In response to the challenge that no scientist has ever seen evolution occurring, Richard Dawkins shares that nobody has ever seen it happening:

◀ *We observe evolution - we just have never seen evolution while it is happening.* ▶ [22]

There is no credible evidence that dinosaurs evolved into birds. Evolutionary scientists promote the now extinct bird Archaeopteryx as the supposed link between dinosaurs and birds. If dinosaurs really evolved into birds, because of the incredible anatomical and physiological differences between dinosaurs and birds - such as bone densities, bone structures, circulation systems, and feathers (to name only a few) - there would be thousands if not millions of transitional dinosaur-to-bird fossils.

No new advancements or scientific discoveries, in any field, have ever been based on evolution. Ponder that for a moment... if evolution was true, there would be hundreds if not thousands of scientific discoveries directly associated or related to an aspect of evolution. Paleontologist and evolutionist Colin Patterson, Ph.D., who worked at the London British Museum of Natural History, addressed a group of evolutionists in a 1981 speech at the American Museum of Natural History in New York, and shares:

◀ *One morning I woke up and something had happened in the night, and it struck me that I had been working on this stuff for 20 years and there was not one thing I knew about it. That's quite a shock to learn that one can be misled so long. Either there was something wrong with me, or there was something wrong with evolution theory... [He began to ask his peers] Can you tell me anything you know about evolution, any one thing that is true? I tried that question on the geology staff at the Field Museum of Natural History, and the only answer I got was silence. [He asked the Evolutionary Morphology Seminar at the University of Chicago and all he received] ...was silence for a long time and eventually one person said, 'I do know one thing - it ought not to be taught in high school.' ... It does seem that the level of knowledge about evolution is remarkably shallow. We know it ought not to be taught in high school, and that's all we know about it... Then I woke up and realized that all my life I had been duped into taking evolution as revealed truth in some way... I feel that the effects of hypotheses of common ancestry in systematics has not been merely boring, not just a lack of knowledge, I think it has been positively anti-knowledge.* ▶ [23]

The famous Stanley Miller chemistry experiment, still referenced in middle school, high school, and college science books to support the idea that life can arise from non-life, actually has unsolvable scientific problems with its results, such that it does not offer scientific support to evolution. Geneticist and creationist Jeffrey Tomkins, Ph.D., identifies four unresolvable scientific problems with the Stanley Miller experiment results: [24]

1. The amino acids produced in this experiment were only the simplest forms, not the complex forms that living cells actually build.

2. The amino acids produced in this experiment included both "left-handed" and "right-handed" versions, but living cells only use left-handed forms. Any primordial soup would offer no way to exclude all right-handed amino acids and cause only left-handed amino acids to bond exclusively together in order to form proteins.

3. The experiment actually produced a caustic tar-like goo which contained chemicals that destroy living cells, including hydrogen cyanide and formaldehyde.

4. The experiment destroyed amino acids as fast as it would create them.

Biologist and creationist Dean H. Kenyon, Ph.D., states in an affidavit to the U.S. Supreme Court that evolutionary geneticists have been unable to provide any experimental research in which left-handed amino acids only bond with other left-handed amino acids:

◀ *Many researchers have attempted to find plausible natural conditions under which L-amino acids [left-handed] would preferentially accumulate over their D-counterparts [right-handed], but all such attempts have failed. Until this crucial problem is solved, no one can say that we have found a naturalistic explanation for the origin of life. Instead, these isomer preferences point to biochemical creation.* ▶ [25]

Not all scientists believe that evolution is a fact. Physiologist N. J. Tahmisian, Ph.D., strongly speaks out against evolution being a fact:

◀ *Scientists who go about teaching that evolution is a fact of life are great con-men, and the story they are telling may be the greatest hoax ever. In explaining evolution, we do not have one iota of fact.* ▶ [26]

M.I.T. mathematician and physicist Wolfgang Smith, Ph.D., points out that evolution actually does not have any legitimate and conclusive scientific evidence:

◀ *Evolution is totally bereft of scientific sanction. There exists to this day not a shred of bona fide scientific evidence in support of the thesis that macroevolutionary transformations have ever occurred... The point, however, is that the doctrine of evolution has swept the world, not on the strength of its scientific merits, but precisely in its capacity as a Gnostic myth. It affirms, in effect, that living beings created themselves, which is, in essence, a metaphysical claim... Thus, in the final analysis, evolutionism is in truth a metaphysical doctrine decked out in scientific garb.* ▶ [27]

Biochemist and agnostic Michael Denton, Ph.D., in his book *Evolution: A Theory in Crisis* pinpoints that Charles Darwin, the recognized father of evolution, actually could not scientifically identify even one new species that was the result of natural selection:

◀ *It was not only his [Charles Darwin] general theory that was almost entirely lacking in any direct empirical support, but his special theory was also largely dependent on circumstantial evidence. A striking witness to this is the fact that nowhere was Darwin able to point to one bona fide case of natural selection having actually generated evolutionary change in nature, let alone having been responsible for the creation of a new species.* ▶ [28]

Biochemist David Ezra Green, Ph.D., from the Institute of Enzyme Research, and chemist Robert F. Goldberger, Ph.D., of the U.S. National Institutes of Health, point out that evolution scientifically fails to be a testable hypothesis and is instead merely conjecture:

◀ *The macromolecule-to-cell transition is a jump of fantastic dimensions, which lies beyond the range of testable hypothesis. In this area all is conjecture.* ▶ [29]

The scientific community as a whole has changed the proposed age of the earth over 100 times in the last couple centuries, which certainly does not speak to a high level of scientific certainty. Geologist and evolutionist A. E. J. Engel notes the many accepted changes by the scientific community for the earth's age over the last couple centuries:

◀ *The fact that the calculated age of the earth has increased by a factor of roughly 100 between the year 1900 and today - as the accepted 'age' of the earth has increased from about 50 million years in 1900 to at least 4.6 aeons [billions] today - certainly suggests we clothe our current conclusions regarding time and the earth with humility.* ▶ [30]

In a *Newsweek* article titled "Science Contra Darwin", reporter Sharon Begley shares the following insights about the scientific integrity of evolution:

◀ *The great body of work derived from Charles Darwin's revolutionary 1859 book, On the Origin of Species, is under increasing attack - and not just from creationists... So heated is the debate that one Darwinian says there are times when he thinks about going into a field with more intellectual honesty: the used car business.* ▶ [31]

Scientific Reason 77 Summary: No conclusive scientific evidence exists to support evolution over billions of years.

Chapter 10 Endnotes

1 – Geoff Downes, *In Six Days – Why 50 Scientists Choose to Believe in Creation*, Editor John F. Ashton (Green Forest, AR: Master Books, 2001) page 334.

2 – Scott C. Todd, "Wellcome for education on science in society", *Nature*, Vol. 401, No. 6,752 (30September1999) page 423.

3 – Isaac Newton, *Goodreads.com*, Web 11January2016. <https://www.goodreads.com/author/quotes/135106.Isaac_Newton>

4 – Fred Hoyle, "The Big Bang in Astronomy", *New Scientist*, Vol. 92 (19November1981) pages 526-527.

5 – Edwin Conklin, *Reader's Digest* (January 1963) page 92.

6 – William Dembski, "Who's Got the Magic?", Designinference.com (25April2000) Web 21September2015. <http://designinference.com/documents/2000.04.pennock_response.htm>

7 – I. L. Cohen, *Darwin Was Wrong: A Study in Probabilities* (New York: New Research Publications, Inc., 1984) page 81.

8 – Stephen Jay Gould, "Not Necessarily a Wing", *Natural History*, Vol. 94 (October 1985) page 13.

9 – Georgia Purdom, *The New Answers Book 1*, Editor Ken Ham (Green Forest, AR: Master Books, 2010) page 281.

10 – Walter J. Bock, "Evolution by Orderly Law", *Science*, Vol. 164 (9May1969) pages 684-685.

11 – Jeremy Rifkin, *Algeny* (New York: Viking Press, 1983) page 145.

12 – George Gaylord Simpson and William Beck, *An Introduction to Biology* (New York: Harcourt Brace and World, 1965) page 241.

13 – Keith S. Thompson, "Ontogeny and Phylogeny Recapitulated", *American Scientist*, Vol. 76 (May/June 1988) page 273.

14 – David Catchpoole and Jonathan Sarfati, "Schweitzer's Dangerous Discovery", *Creation.com*, Pub 21July2014. Web 21September2015. <http://creation.com/schweitzers-dangerous-discovery>

15 – Robert Kofahl, "A Serious Problem for Homology", *Answers in Genesis*, Answersingenesis.org, Pub 1March1992. Web 4January2016. <https://answersingenesis.org/biology/a-serious-problem-for-homology>

16 – Gunter Wagner, "The origin of morphological characters and the biological basis of homology", *Evolution*, Vol. 6, No. 43 (1989) page 1,163.

17 – Jerry Bergman, "Does homology provide evidence of evolutionary naturalism?", *Journal of Creation*, Vol. 15, Iss. 1 (April 2001) pages 26-33. <http://creation.com/does-homology-provide-evidence-of-evolutionary-naturalism>

18 – Michael R. Rose, "Darwin was Wrong", *New Scientist* (21January2009). <https://www.newscientist.com/article/mg20126921-600-why-darwin-was-wrong-about-the-tree-of-life>

19 – Jerald M. Loewenstein and Adrienne L. Zihlman, "The Invisible Ape", *New Scientist*, Vol. 120 (3December1988) page 58.

20 – Colin Patterson, “Cladistics”, *Interview on BBC TV by Peter Franz*, Producer Brian Leek (4March1982).
21 – Frank Sherwin, “Human Evolution An Update”, *Acts & Facts*, Vol. 9 (1997) page 26. <http://www.icr.org/article/human-evolution-update>
22 – Bill Moyers, “NOW with Bill Moyers”, *Moyers & Company*, BillMoyer.com, Pub 1March2013. Web 21September2015. <http://billmoyers.com/2013/03/01/moyers-moment-2004-richard-dawkins-on-the-truth-of-evolution>
23 – Colin Patterson, "Evolution and Creationism", *Transcript of Speech at American Museum of Natural History* (5November1981).
24 – Jeffrey Tomkins, *Creation Basics and Beyond*, Editor Jason Lisle (Dallas, TX: Institute For Creation Research, 2013) pages 188-189.
25 – Dean H. Kenyon, “Brief of Appellants”, *Affidavit presented to U.S. Supreme Court*, No. 85-15, 13, prepared under the direction of William J. Guste, Jr., Attorney General of the State of Louisiana (October 1985) page A-23.
26 – N. J. Mitchell, *Evolution and the Emperor's New Clothes* (U.K.: Roydon Publications, 1983) quoting Dr. T. N. Tahmisian of the Atomic Energy Commission, U.S.A., as stated in the "The Fresno Bee" (20August1959).
27 – Wolfgang Smith, *Teilhardism and the New Religion* (Rockford, IL: Tan Books, 1988) pages 5, 6 and 242.
28 – Michael Denton, *Evolution: A Theory in Crisis* (London: Burnett Books, Limited, 1985) page 62.
29 – David Ezra Green and Robert F. Goldberger, *Molecular Insights in the Living Process* (New York: Academic Press, 1967) page 407.
30 – A. E. J. Engel, “Time and the Earth”, *American Scientist*, Vol. 57 (1969) page 461.
31 – Sharon Begley, “Science Contra Darwin”, *Newsweek* (8April1985) page 80.

Chapter 10 Resources

Naturalistic Assumption

- http://creation.com/assumptions-in-science
- https://answersingenesis.org/what-is-science/what-is-science

Statistical Impossibility

- http://www.icr.org/article/mathematical-impossibility-evolution
- http://designinference.com/documents/2000.04.pennock_response.htm
- *Evolution's Achilles' Heel* by Creation Ministries International
- *In Six Days: Why 50 Scientists Believe in Creation: Chapter 2* by Master Books

Natural Selection

- http://creation.com/natural-selection-questions-and-answers
- https://answersingenesis.org/natural-selection/natural-selection-vs-evolution
- https://answersingenesis.org/theistic-evolution/jacques-monod-and-theistic-evolution
- http://www.icr.org/article/darwins-sacred-imposter-illusion-that
- *The New Answers Book 1: Chapter 22* by Answers in Genesis
- *Evolution's Achilles Heel: Chapter 1* by Creation Ministries International
- *Evolution the Grand Experiment: Chapter 4* by Carl Werner

Speciation

- https://answersingenesis.org/natural-selection/speciation
- http://creation.com/speciation-questions-and-answers

Haeckel's Biogenetic Law

- http://creation.com/ernst-haeckel-evangelist-for-evolution-and-apostle-of-deceit
- http://creation.com/embryo-design
- http://www.icr.org/article/heritage-recapitulation-theory
- http://www.creationscience.com/onlinebook/ReferencesandNotes20.html

Extinction of Dinosaurs

- http://www.icr.org/soft-tissue-list (Soft Tissue Discoveries)
- http://creation.com/dinosaur-questions-and-answers
- http://creation.com/schweitzers-dangerous-discovery
- *Guide to Dinosaurs* by Institute For Creation Research
- *The New Answers Book 1: Chapter 12* by Answers in Genesis
- *The Creation Answers Book: Chapter 19* by Creation Ministries International
- *Creation Basics and Beyond: Chapter 39* by Institute For Creation Research

- 🕮 *Untold Secrets of Planet Earth – Dire Dragons* by Vance Nelson
- 🕮 *Dragons: Legends and Lore of Dinosaurs* by Master Books

Homologous Structures

- 💻 http://creation.com/homology-made-simple
- 💻 http://creation.com/not-to-be-used-again-homologous-structures-and-the-presumption-of-originality-as-a-critical-value
- 💻 http://creation.com/does-homology-provide-evidence-of-evolutionary-naturalism
- 💻 http://www.icr.org/article/does-homology-prove-evolution
- 💻 https://answersingenesis.org/biology/a-serious-problem-for-homology
- 💻 https://answersingenesis.org/theory-of-evolution/evidence/similarities-dont-prove-evolution
- 💻 https://answersingenesis.org/biology/comparative-similarities-homology
- 🕮 *Evolution the Grand Experiment: Chapter 5* by Carl Werner
- 🕮 *Creation Basics and Beyond: Chapter 18* by Institute For Creation Research

Morphological Tree Diagrams

- 💻 http://creation.com/cladistics
- 💻 https://answersingenesis.org/theory-of-evolution/evolution-timeline/open-tree-life-summarizes-evolutionary-beliefs
- 💻 https://www.newscientist.com/article/mg20126921-600-why-darwin-was-wrong-about-the-tree-of-life
- 🕮 *Creation Basics and Beyond: Chapter 20* by Institute For Creation Research

No Conclusive Evidence

- 💻 http://www.icr.org/article/vanishing-case-for-evolution
- 💻 http://www.icr.org/article/391
- 💻 http://thegrandexperiment.com
- 💻 http://billmoyers.com/2013/03/01/moyers-moment-2004-richard-dawkins-on-the-truth-of-evolution
- 🕮 *Evolution the Grand Experiment* by Carl Werner
- 🕮 *The New Answers Book 1: Chapter 24* by Answers in Genesis
- 🕮 *The New Answers Book 2: Chapter 6* by Answers in Genesis

⌘ Chapter 11 ⌘

Biblical Reasons

What all this means is that Christianity cannot lose the Genesis account of creation like it could lose the doctrine of geocentrism and get along. The battle must be waged, for Christianity is fighting for its very life. Christianity has fought, still fights, and will fight science to the desperate end over evolution, because evolution destroys utterly and finally the very reason Jesus' earthly life was supposedly made necessary. Destroy Adam and Eve and the original sin, and in the rubble you will find the sorry remains of the son of god. Take away the meaning of his death. If Jesus was not the redeemer who died for our sins, and this is what evolution means, then Christianity is nothing! [8]

⌘ Biblical Reason 1 ⌘
Order of Creation Events

The following table highlights the historical record listed in Genesis chapters one to three and the contrary events promoted by evolutionary beliefs.

GENESIS CHAPTERS 1-3	EVOLUTIONARY EVENTS
God created matter in the beginning.	Matter existed on its own.
The earth was created before our sun and stars.	Our sun and the stars existed before the earth.
The earth was originally fully covered with water.	The earth was originally a molten lava blob without water.
Oceans existed first, then dry land appeared.	Dry land existed first, then the oceans appeared.
Life was first created on land.	Life started in the oceans.
Fruit trees were created before fish.	Fish existed before fruit trees.
Plants were created before the sun.	Plants developed after our sun existed.
Birds were created before land reptiles.	Land reptiles existed before birds.
Birds were created before land animals.	Land animals existed before birds.
Whales were created before land animals.	Land animals existed before whales.
Adam was created first, then Eve was created from Adam.	A female existed genetically before a male existed.
Adam named many of the original created kinds of animals, which included animals we now call dinosaurs.	Many animals, including those we have named dinosaurs, went extinct before people existed.
God finished creating everything in six days and rested on the seventh day.	"Creative" processes are still occurring today.
Adam and Eve were created in a perfect world without any death or struggle, then they sinned and caused death.	Death and struggle have always existed, and people are the end-result of billions of years of death and struggle.

Biblical Reason 1 Summary: The order of events promoted by evolution are completely contrary to the historical order of events detailed in Genesis chapters one to three.

⌘ Biblical Reason 2 ⌘ Comparing Worldviews

Everyone has a worldview. Each person's worldview is simply the way he or she sees the world, what preconceived values and ideas are held, and what is considered reality versus imagination.

Both the Bible and evolution promote a worldview. The following table contrasts the Biblical worldview, of people and life itself, with the worldview promoted by evolution.

BIBLICAL WORLDVIEW	EVOLUTIONARY WORLDVIEW
People are purposefully created in the image of the triune God.	People are merely the result of random and purposeless chances.
People are the primary object of Jesus' love here on earth.	People are just advanced forms of evolving animals struggling in a "survival of the fittest" world.
People are eternal.	People are only temporal.
Every person is a special, unique creation of the Creator Jesus Christ.	Every person is only the result of random chemical processes that just happened to align over billions of years.
People are to be stewards of the Lord's creation.	People are the most advanced animal on the earth because they have "evolved" more than other life forms.
Life has meaning now and into eternity.	Life has no meaning and existence ends with death.
The Bible speaks of a future hope, in which death, suffering, disease, and tears will no longer exist.	Evolution offers no future hope, only a world that will fizzle out in a heat death when usable energy is depleted.
The Lord Jesus Christ offers a peace both now on earth and in the future with Him.	Evolution offers no real peace, and people must continually struggle and fight to survive.
Life has value and is precious.	Life has no dignity or value - the strong can and should conquer the weak.
People are the focus of God Almighty's plan of redemption.	When someone dies, they are done and gone.
God the Father, God the Son, and God the Holy Spirit are transcendent throughout all time and history.	God the Father, God the Son, and God the Holy Spirit do not exist.
The universe and everything in it is ordered by the Lord Jesus Christ for His purposes.	The universe is the result of an initial explosion, and any appearance of order is just luck.

Mankind's primary purpose is to glorify the Lord God Almighty and live in daily fellowship with the Lord Jesus.	Struggle, survival, and ultimately death are man's primary purposes.
Creation, even under the curse of sin, displays beautiful intricacy, order, and design.	Things around us are merely the result of an initial chaotic explosion, and any appearance of order is just coincidence.
There is an infinite and purposeful Lord Jesus Christ.	There are an infinite number of random chances.
Life always and only comes from other life, and death is an enemy to life.	Life originally came from non-life, and today death causes more advancing and evolving forms of life.
The Triune God is the creator.	Time and death are the creators.
Entropy is occurring all around us.	Improvement is occurring all around us.
The landscapes we see on earth are the result of catastrophic effects from Noah's global flood completely covering and destroying the earth about 4,300 years ago.	The landscapes we see on earth formed gradually and slowly over billions of years.
Absolutes exist.	Everything is relative to an individual's personal perspective.
The age of the universe and our earth is about 6,000 years old.	The universe is about 15 billion years old, and our earth is about 4.5 billion years old.
There have been intelligent civilizations from the start of human history about 6,000 years ago.	People have survived in primitive and nomadic conditions for millions of years, with intelligent civilizations only existing during the last few thousand years.

Chemist Frank Sherwood Taylor, Ph.D., was a British historian of science who served as the Director of the Science Museum in London. On the subject of worldviews, he offers a chilling reflection about what belief in evolution's worldview can do when a culture embraces it:

◀ *I myself have little doubt that in England it was uniformitarian, long-ages geology and the theory of evolution that changed us from a Christian to a pagan nation.* ▶ [1]

Biochemist and agnostic Michael Denton, Ph.D., shares the catastrophic and destructive impact that belief in evolution over billions of years has had on the Biblical worldview:

◀ *As far as Christianity was concerned, the advent of the theory of evolution and the elimination of traditional teleological thinking was catastrophic. The suggestion that life and man are the result of chance is incompatible with the Biblical assertion of their being the direct result of intelligent creative activity. Despite the attempt by liberal theology to disguise the point, the fact is that no Biblically derived religion can really be compromised with the fundamental assertion of Darwinian theory. Chance and design are antithetical concepts, and the decline in religious belief can probably be attributed more to the propagation and advocacy by the intellectual and scientific community of the Darwinian version of evolution than to any other single factor.* ▶ [2]

Philosopher James Rachels, Ph.D., notes that evolutionary beliefs directly attack the dignity of human life:

◀ *Darwinism undermines both the idea that man is made in the image of God and the idea that man is a unique rational being. Furthermore, if Darwinism is correct, it is unlikely that any other support for the idea of human dignity will be found.* ▶ [3]

During an interview with John Laurie of the Australian Broadcasting Company, biologist and evolutionist Jacques Monod, Ph.D., highlights the cruelty inherent within the idea of natural selection:

◀ *[Natural] selection is the blindest and most cruel way of evolving new species and more and more complex and refined organisms... The struggle for life and the elimination of the weakest is a horrible process, against which our whole modern ethics revolts. An ideal society is a non-selective society, one where the weak is protected; which is exactly the reverse of the so-called natural law. I am surprised that a Christian would defend the idea that this is the process which God more or less set up in order to have evolution.* ▶ [4]

Biologist and evolutionist Sir Julian Sorell Huxley, whose grandfather Thomas Huxley was a close friend of Charles Darwin, identifies that belief in evolution is not compatible with a Biblical worldview:

◀ *Darwinism removed the whole idea of God as the creator of organisms from the sphere of rational discussion. Darwin pointed out that no supernatural designer was needed; since natural selection could account for any known form of life, there was no room for a supernatural agency in its evolution... I think we can dismiss entirely all ideas of a supernatural overriding mind being responsible for the evolutionary process.* ▶ [5]

Atheist and evolutionist Frank Zindler understands that evolution cannot be harmonized with a Biblical worldview:

◀ *The most devastating thing though that biology did to Christianity was the discovery of biological evolution. Now that we know that Adam and Eve never were real people, the central myth of Christianity is destroyed. If there never was an Adam and Eve, there never was an original sin. If there never was an original sin, there is no need of salvation. If there is no need of salvation, there is no need of a saviour. And I submit that puts Jesus, historical or otherwise, into the ranks of the unemployed. I think that evolution is absolutely the death knell of Christianity.* ▶ [6]

Sociologist and evolutionist Jeremy Rifkin certainly understands the distinction between the Biblical worldview and an evolutionary worldview:

◀ *We no longer feel ourselves to be guests in someone else's home and therefore obliged to make our behavior conform with a set of preexisting cosmic rules. It is our creation now. We make the rules. We establish the parameters of reality. We create the world, and because we do, we no longer feel beholden to outside forces. We no longer have to justify our behavior, for we are now the architects of the universe. We are responsible to nothing outside ourselves, for we are the kingdom, the power, and the glory for ever and ever.* ▶ [7]

Biblical Reason 2 Summary: The worldview of the Bible and the worldview promoted by evolution are completely incompatible.

⌘ Biblical Reason 3 ⌘
Foundation Attack

Ultimately, every single Biblical doctrine of theology, directly or indirectly, is founded in the historical account given in Genesis chapters 1 to 11. Reinterpreting Genesis history according to non-Biblical beliefs, and holding to a position that everything created is somehow the result of evolution over billions of years, undermines the foundation for the rest of the Bible, and ultimately the need for the gospel of grace and forgiveness freely offered by the Lord Jesus Christ!

Interpreting the seven days of the creation week as long ages (millions or billions of years) brings outside ideas and interpretations into the Bible (eisegesis) rather than letting scripture interpret scripture and receiving the clear grammatical and historical understanding that the verses offer (exegesis). Tragically, atheist G. Richard Bozarth understands this better than many people do:

◀ *What all this means is that Christianity cannot lose the Genesis account of creation like it could lose the doctrine of geocentrism and get along. The battle must be waged, for Christianity is fighting for its very life. Christianity has fought, still fights, and will fight science to the desperate end over evolution, because evolution destroys utterly and finally the very reason Jesus' earthly life was supposedly made necessary. Destroy Adam and Eve and the original sin, and in the rubble you will find the sorry remains of the son of god. Take away the meaning of his death. If Jesus was not the redeemer who died for our sins, and this is what evolution means, then Christianity is nothing!* ▶ [8]

Josef Ton, a Romanian Baptist pastor imprisoned for his faith under the Romanian communist government in the twentieth century, identifies why Christianity is not compatible with evolution and its supposed billions of years of earth's history:

◀ *I came to the conclusion that there are two factors which destroyed Christianity in Western Europe. One was the theory of evolution, the other, liberal theology... Liberal theology is just evolution applied to the Bible and our faith.* ▶ [9]

Theologian G. H. Pember identifies the danger of trying to harmonize the word of God (the Bible) with evolution and its billions of years of earth's history. His powerful quote reminds us that evolution attacks the foundation of Genesis, leaving in its wake people who no longer know how to discern the truth of the Bible from the teachings of men:

◀ *For, by skillfully blending their own systems with the truths of scripture, they so bewildered the minds of the multitude that but few retained the power of distinguishing the revelation of God from the craftily interwoven teachings of men.* ▶ [10]

Biblical Reason 3 Summary: Evolution attacks the very foundation of the Bible, which is provided in the book of Genesis starting at chapter one.

⌘ Biblical Reason 4 ⌘ Naturalism's Failures

Meteorologist and creationist George S. Hawke, Ph.D., identifies that the belief in naturalism, which is based upon the idea that only materialistic or naturalistic forces (those based solely on matter) caused everything around us to exist, fails three "tests": the Jesus test, the Gospel test, and the Fruit test.

Evolution is based upon the belief of naturalism, and thus, it excludes the need for a creator or anything supernatural. The Bible clearly teaches that the Lord Jesus created and sustains everything in our universe, and of course, He is a supernatural being. Thus, naturalism contradicts the Bible and the existence of the Lord Jesus.

Naturalism is based upon the belief that only natural forces can be given credit for the origin of our universe and everything within it, including people. It denies anything in the Bible, from creation to the reality of sin and our need for a Savior. Thus, naturalism contradicts the Bible and its gospel message of the need for a Savior for every person.

The belief in naturalism supports the philosophies of materialism (striving for material things), humanism (people make the rules and are the highest authority in life), and pantheism (nature replaces the Lord Jesus Christ as ruler of the universe). The Bible says we can discern the value of a belief by the fruit it produces, and naturalism produces spiritual death and destruction. Thus, naturalism contradicts the Bible and produces fruit that is rotten and destructive. Dr. Hawke summarizes why the philosophy of naturalism contradicts the clear teachings of the Bible:

◀ *It is clear from this that the viewpoint of naturalism fails all three Biblical tests for determining what is true. Therefore, it is false and is not consistent with the overall message of the Bible.* ▶ [11]

Biblical Reason 4 Summary: The belief system of naturalism is a foundational philosophy for evolution that is completely incompatible with the Bible.

⌘ Biblical Reason 5 ⌘ Theistic Evolution

Evolution and the attributes of the three members of the Trinity are completely incompatible. In an attempt to harmonize the Bible with evolution over billions of years, it has been suggested that God used evolution in order to perform His supernatural act of creation. This suggestion has been coined "theistic evolution".

Evolution promotes that random, purposeless, materialistic, naturalistic, unintelligent, goalless, death-based, designless, chance processes are responsible for the origin and continuation of everything around us, including material things, spiritual things, and life itself. The attributes of God the Father are exactly the opposite: He is orderly and not random. He possesses perfect knowledge, intelligence, understanding and wisdom. He supernaturally created all matter and life and now continually sustains all life on earth. The Lord Jesus Christ is not only called the Word but His word models a perfect purpose, which directly contradicts evolutionary random chance processes, as emphasized in Isaiah 55:11:

◀ *So shall My word be that goes out from My mouth; it shall not return to Me empty, but it shall accomplish that which I purpose, and shall succeed in the thing for which I sent it.* ▶

It is not Biblical, in any sense, to believe that God used evolution to accomplish His supernatural act of creation, nor is it Biblical in any sense to believe He needed billions of years to complete His creation. Ironically, atheist and evolutionist G. Richard Bozarth understands this better than any theistic evolutionist does:

◀ *Christianity is - must be! - totally committed to the special creation as described in Genesis, and Christianity must fight with its full might, fair or foul, against the theory of evolution. It becomes clear now that the whole justification of Jesus' life and death is predicated on the existence of Adam and the forbidden fruit he and Eve ate. Without the original sin, who needs to be redeemed? Without Adam's fall into a life of constant sin terminated by death, what purpose is there to Christianity? None.* ▶ [12]

Biblical Reason 5 Summary: The proposal of theistic evolution and the attributes of God the Father, God the Son, and God the Holy Spirit are completely incompatible.

⌘ Biblical Reason 6 ⌘
Gap Theory

The gap theory proposes that the Bible teaches a gap of millions or billions of years occurring between verses one and two in Genesis chapter one. Genesis chapter one, verses one and two state:

◀ *1: In the beginning, God created the heavens and the earth. 2: The earth was without form and void, and darkness was over the face of the deep. And the Spirit of God was hovering over the face of the waters.* ▶

A clear, straightforward reading of these verses offers no gap of any time - they merely state that God created the heavens and the earth, and the originally created earth was without form and void.

The gap theory is not supportable with the verses in Genesis chapter one or with any other Old Testament or New Testament verses. The verses of Genesis chapter one teach us that God created all things during the seven days of the creation week, and there is no reference to a gap of millions or billions of years of time between the first two verses nor between any of the verses describing the days of creation.

For the gap theory to be true, the days of Genesis would have to include much more time than just seven days. Reasonably and logically, only one understanding can be true: the creation week occurred in only seven days or the creation week included millions or billions of years in a gap of time between verses one and two.

Biblical Reason 6 Summary: A clear, straightforward reading of Genesis chapters one and two teaches us that the gap theory and the clear context and meaning communicated in the first two chapters of Genesis are not compatible.

⌘ Biblical Reason 7 ⌘
Writing Style of Genesis

Some people believe that Genesis supports billions of years of time, claiming that the majority of Genesis, especially chapters one to three, are written as poetry, parable, or allegory. Actually, the first three chapters of Genesis are written as historical narrative, and the historical and grammatical style that Moses used to write these chapters describes real history and events.

Hebrew professor James Barr, Ph.D., who taught at Oxford University in England as well as Vanderbilt University, identifies that the Hebrew writing style of Genesis is not compatible with evolution and its billions of years. In 1984, Dr. Barr, who himself did not believe Genesis to be actual history, acknowledged that as far as the language of Genesis chapters 1 to 11 is written:

◀ *Probably, so far as I know, there is no professor of Hebrew or Old Testament at any world-class university who does not believe that the writer(s) of Genesis 1 to 11 intended to convey to their readers the ideas that (a) creation took place in a series of six days which were the same as the days of 24 hours we now experience (b) the figures contained in the Genesis genealogies provided by simple addition a chronology from the beginning of the world up to later stages in the Biblical story (c) Noah's flood was understood to be world-wide and extinguish all human and animal life except for those in the ark. Or, to put it negatively, the apologetic arguments which suppose the 'days' of creation to be long eras of time, the figures of years not to be chronological, and the flood to be a merely local Mesopotamian flood, are not taken seriously by any such professors, as far as I know.* ▶ [13]

Notice why Dr. Barr's quote is so significant: he himself does not believe in a historical, seven-day creation week, but rather, he is what is referred to as a "hostile" witness to these chapters of Genesis. And that provides the power of his quote: he recognizes that the author of Genesis, which we know was Moses writing under the inspiration of the Holy Spirit, is conveying that the days of the creation week were actual days, not shorter timeframes such as seconds nor longer timeframes such as billions of years.

Edward J. Young, professor of Hebrew at Westminster Seminary in Philadelphia, also identifies that the writing style of Genesis is not compatible with evolution and its billions of years:

◀ *To escape from the plain factual statements of Genesis some evangelicals are saying that the early chapters of Genesis are poetry or myth by which they mean that they are not to be taken as straightforward accounts and that the acceptance of such a view removes the difficulties. To adopt such a view, they say, removes all troubles with modern science. Genesis is not poetry. There are more poetical accounts of creation in the Bible, such as Psalm 104, certain chapters of Job, and they differ completely from the first chapter of Genesis. Hebrew poetry has certain characteristics and they are not found in the first chapter of Genesis. So the claim that Genesis one is poetry is no solution at all. The man who says, 'I believe Genesis purports to be a historical account, but I do not believe that account', is a far-better interpreter of the Bible than the man who says, 'I believe Genesis is profoundly true, but it is poetry.'* ▶ [14]

Biblical Reason 7 Summary: The writing style of Genesis is a historical narrative that is not written as poetry or as a parable or as an allegory. Therefore, the seven-day creation week described in the first two chapters of Genesis is not compatible with evolution and its required billions of years of earth's history.

⌘ Biblical Reason 8 ⌘ Clear Meaning of Genesis Chapters 1 and 2

Genesis chapters one and two describe God's creation of all physical material and life itself over six days. Observing how the Hebrew word Yom (day) is used in these chapters provides clear meaning and understanding of the verses. [15]

In each of the six days of creation, the Hebrew word Yom is modified with a number. We read references to the first day, the second day, and continuing up through the sixth day. Everywhere in the Old Testament, when Yom is modified with a number, it means a literal 24 hour day. Outside of Genesis chapter one, considering both the Old Testament and New Testament, Yom is used with a number over 350 times, and each time it means an ordinary day. It is inconsistent to think that Genesis chapter one is the exception to all scripture regarding its use of the word Yom when modified with a number.

In Genesis 1:5, Yom is used with "night". Outside of Genesis chapter one, "night" is used with Yom over 50 times, and each time it means a literal day. In this verse, the reference to light and night together means a literal 24 hour day. If this day represented billions of years, it would make no sense to suggest that the light portion of the day was hundreds of millions of years long and the night portion of the day was also hundreds of millions of years long. Genesis 1:5 reads:

◀ *God called the light Day, and the darkness He called Night. And there was evening and there was morning, the first day.* ▶

A number and the phrase "evening and morning" are used to identify and distinguish each of the six days of creation described in Genesis chapter one. Outside of Genesis chapter one, the Hebrew word Yom is used individually with the word "evening" or individually with the word "morning" over 20 times. "Evening" and "morning" appear together, but without the word Yom, over 30 times. In all of these times the context refers to an ordinary 24 hour day. It only makes sense to reference the evening of each day or the morning of each day if the six days of creation are each actual days. Additionally, it makes no logical sense to refer to the morning of a million or billion years, or the evening of a million or billion years.

Hebrew possesses a plural form or conjugation of the word Yom, which can be used for expressions such as "in those days". This plural usage of the word Yom is not used in Genesis chapter one. There are also Hebrew words Olam and Qedem that are used to communicate long periods of time, or even indefinite amounts of time, but neither of these words are used in Genesis chapter one.

Genesis 1:14-19 describes some of the Lord's creation activity on the fourth day of the creation week. These verses use a meaning of the word day describing the light portion of a normal day. The verses also use a meaning of the word day describing a 24 hour period of time. With no intent to be blasphemous with God's word, but in order to further emphasize the point that Genesis chapter one verses do not support any use of the word day meaning billions of years, verses 14-19 below have been modified with all uses of the word day replaced with "billions of years" to show how silly this is:

◀ *And God said, Let there be lights in the firmament of the heaven to divide the [billions of years] from the night; and let them be for signs, and for seasons, and for [billions of years], and years: And let them be for lights in the firmament of the heaven to give light upon the earth: and it was so. And God made two great lights; the greater light to rule the [billions of years], and the lesser light to rule the night: he made the stars also. And God set them in the firmament of the heaven to give light upon the earth, and to rule over the [billions of years] and over the night, and to divide the light from the darkness: and God saw that it was good. And the evening and the morning were the fourth [billions of years].* ▶

Some people have proposed that perhaps God created only the earth and our local solar system during the six days of creation, but not the entire universe and everything else within it. Genesis chapter one states that God created the "heavens and the earth". The expression "heavens and the earth" demonstrates the usage of a merism, which is a language structure in which opposites are combined to represent an all-encompassing idea. This merism expression "heavens and the earth" represents everything in all of creation, not just a subset of everything created.

A good friend of Charles Darwin, evolutionist Thomas Huxley, who was nicknamed "Darwin's Bulldog", did not believe the verses in the first chapters of Genesis but none-the-less fully understood the significance of what they stated and how they should be understood:

◀ *If divine authority is not here claimed for the twenty-fourth verse of the second chapter of Genesis, what is the value of language? And again, I ask, if one may play fast and loose with the story of the Fall as a 'type' or 'allegory', what becomes of the foundation of Pauline theology? The melancholy fact remains, that the position they have taken up is hopelessly untenable.* ▶ [16]

Biblical Reason 8 Summary: The straightforward meaning and clear contextual usage of the Hebrew word Yom used in Genesis chapters one and two teach us that the seven days of the Biblical creation week are completely incompatible with evolution and its billions of years of earth history.

⌘ Biblical Reason 9 ⌘
View of People

The Bible claims people are made in the image of God Himself. In contrast, evolution claims people are the result of struggle, death, and billions of genetic and physical random-chance coincidences miraculously occurring over billions of years.

The Biblical view distinguishes people as the most special creation in everything the Lord Jesus Christ created. Of everything created during the creation week, including the earth, the other planets, the stars, and all the kinds of plants, fish, birds and animals, people have been created uniquely. There are several ways we know people have been created uniquely compared to everything else that has been created. Of all creation, people are the only creatures:

- created in the image of God
- uniquely bipedal (See below)
- able to fully express emotions, such as joy, sadness, hope, and laughter
- that received the breath of God when created
- for which God used His hands to create
- that can actually speak with God
- that possess free will
- that have the ability to think creatively and artistically
- created with an eternal soul

When it comes to being bipedal, people are unique compared to all other creatures. This uniqueness involves the way the shapes, angles, and relationships of our bones, muscles, and connecting tissues function together within our feet, ankles, knees, calves, thighs, hips, spine, arms, and shoulders. [17]

Biblical Reason 9 Summary: The Biblical view of the unique and special creation of people is completely incompatible with evolution's view of people.

⌘ Biblical Reason 10 ⌘
Genesis Genealogies

The genealogies listed in Genesis chapters 5, 7, and 11 help us determine the approximate age of our earth within about 100 years. [18]

The following table is a listing of Bible verses we can use to determine the approximate age of the earth.

- The "Verse" column tells us the Genesis verse where specific events and timeframes are given.
- The "Text of Verse" column provides text from the scripture verse which tells us specific events and timeframes.
- The "Womb Time" column identifies the extra nine months of time from when a man fathered his son to when the son was born.
- The "Amt. of Time" column tells us the amount of time stated in the verse.
- The "Partial Time" column accounts for the extra days of time that are not specifically called out in the verse relative to a father's age. For example, Genesis 5:3 tells us that Adam had lived 130 years when he fathered Seth, but it does not say if he was 130 years and 1 day old or 130 years and 364 days old.
- The "Total Time" column is a total of all time in that row, including the gestation time of the son in his mother's womb, the time specifically mentioned in the verse, and the extra days relative to the father's stated age.

Verse	Text of Verse	Womb Time	Amt. of Time	Partial Time	Total Time
Gen 5:3	When Adam had lived 130 years, he fathered a son… and named him Seth	N/A	130 yrs	≤ 1 yr	≤ 131 yrs
Gen 5:6	When Seth had lived 105 years, he fathered Enosh	≤ 1 yr	105 yrs	≤ 1 yr	≤ 107 yrs
Gen 5:9	When Enosh had lived 90 years, he fathered Kenan	≤ 1 yr	90 yrs	≤ 1 yr	≤ 92 yrs

Gen 5:12	When Kenan had lived 70 years, he fathered Mahalalel	≤ 1 yr	70 yrs	≤ 1 yr	≤ 72 yrs
Gen 5:15	When Mahalalel had lived 65 years, he fathered Jared	≤ 1 yr	65 yrs	≤ 1 yr	≤ 67 yrs
Gen 5:18	When Jared had lived 162 years he fathered Enoch	≤ 1 yr	162 yrs	≤ 1 yr	≤ 164 yrs
Gen 5:21	When Enoch had lived 65 years, he fathered Methuselah	≤ 1 yr	65 yrs	≤ 1 yr	≤ 67 yrs
Gen 5:25	When Methuselah had lived 187 years, he fathered Lamech	≤ 1 yr	187 yrs	< 1 yr	≤ 189 yrs
Gen 5:28	When Lamech had lived 182 years, he fathered a son and called his name Noah	≤ 1 yr	182 yrs	≤ 1 yr	≤ 184 yrs
Gen 7:6	Noah was 600 years old when the flood of waters came upon the earth	N/A	600 yrs	≤ 1 yr	≤ 601 yrs
Gen 11:10	When Shem was 100 years old, he fathered Arpachshad 2 years after the flood	≤ 1 yr	2 yrs	≤ 1 yr	≤ 4 yrs
Gen 11:12	When Arpachshad had lived 35 years, he fathered Shelah	≤ 1 yr	35 yrs	≤ 1 yr	≤ 37 yrs
Gen 11:14	When Shelah had lived 30 years, he fathered Eber	≤ 1 yr	30 yrs	≤ 1 yr	≤ 32 yrs
Gen 11:16	When Eber had lived 34 years, he fathered Peleg	≤ 1 yr	34 yrs	≤ 1 yr	≤ 36 yrs
Gen 11:18	When Peleg had lived 30 years, he fathered Reu	≤ 1 yr	30 yrs	≤ 1 yr	≤ 32 yrs
Gen 11:20	When Reu had lived 32 years, he fathered Serug	≤ 1 yr	32 yrs	≤ 1 yr	≤ 34 yrs
Gen 11:22	When Serug had lived 30 years, he fathered Nahor	≤ 1 yr	30 yrs	≤ 1 yr	≤ 32 yrs
Gen 11:24	When Nahor had lived 29 years, he fathered Terah	≤ 1 yr	29 yrs	≤ 1 yr	≤ 31 yrs
Gen 11:26	When Terah had lived 70 years, he fathered Abram	≤ 2 yr	70 yrs	≤ 1 yr	≤ 72 yrs
			≤ 1,948 yrs		**≤ 1,984 yrs**

The genealogies and events of Genesis chapters 1, 5, 7, and 11 help identify that the creation week happened approximately 6,000 years ago. Here is the math:

- After five days, Adam was created on the sixth day of the creation week.
- There were between 1,948 and 1,984 years from Adam to Abram.
- There were about 2,000 years from Abram (who was later named Abraham) to the birth of Jesus in the city of Bethlehem.
- There have been about 2,016 years from the Jesus' birth to today.

Some people suggest that there are names missing in the Genesis genealogies, and therefore, they cannot be used to determine the approximate age of the earth. This is not a valid argument for three reasons:

1. There would have to be hundreds of thousands of names missing in order for the human lineage to extend back millions of years.

2. Even if there were names not recorded in the genealogies of Genesis chapters 5, 7 and 11, the verses in these genealogy chapters specifically call out the father to son timeframes and a direct lineage. This lineage includes the age of a father when his son was fathered. Using the actual age of the father, and accounting for both gestation time of a son and the extra days in a year for a stated father's age (See above), the lineage adds up to between 1,948 and 1,984 years. So whether names are missing or not, such names would be irrelevant because they would not be part of the direct lineage - rather they would merely be additional descendants whose birthdate and lifespan do not impact the direct lineage. Thus, suggesting that names are missing in the Genesis genealogies is inconsistent and illogical.

3. It does not make sense that the Holy Spirit inspired Moses to write the genealogies, but that He withheld from Moses the names of hundreds of thousands of other direct descendants within the lineage. To be consistent within His perfect attributes, the Holy Spirit cannot lie, or intentionally deceive, or make a mistake. The Holy Spirit has indeed given us the complete genealogy from Adam to Jesus Christ in the gospel of Luke chapter three.

Biblical Reason 10 Summary: The genealogies and timeframes listed in Genesis chapters 1, 5, 7, and 11 help identify that the earth is about 6,000 years old, and therefore, it follows that evolution and its required billions of years of earth's history are not true.

⌘ Biblical Reason 11 ⌘
Created Kinds

Genesis chapter one tells us that creatures and plants were each created after their own kind, a truth directly opposite to the definition of evolution, which promotes only one original common ancestor from which every living thing has descended.

If evolution was true, we would not see the very distinctive lines of biodiversity displayed by the plant and animal kingdom. But in reality, we do see sharp distinctive lines between the kinds of creatures alive today. For example, no one ever confuses a cat with a fish or a hawk. If evolution was true, we would see millions of transitional forms between successive creatures both in the fossil record and also living today. None of these alleged successive creatures, either fossilized or living, exist today.

In Genesis chapter one, 10 times the verses say that plants and animals were created "after their kind". We do not know conclusively if a created kind stated in Genesis one equates to a modern naming category of genus, family, or order, but most creationists think a Genesis kind would be a family distinction within today's binomial naming scheme. Evolution holds that plants and animals descended with modification from one common ancestor that eventually produced all the plants and animals we see living today. The Genesis explanation and evolution's explanation are completely contrary to each other.

Biblical Reason 11 Summary: The idea of one common ancestor from which all life descended, as promoted by evolution over billions of years, is completely incompatible with the Biblical teaching of many distinctly created kinds.

⌘ Biblical Reason 12 ⌘
Thorns

There are fossilized thorns found within sedimentary layers that evolutionary scientists claim are millions of years old and predate man's arrival on earth. [19] In Genesis 3:18, we learn that thorns came into existence because of the curse of sin. Genesis 3:17-19 states:

◀ *And to Adam He said, 'Because you have listened to the voice of your wife and have eaten of the tree of which I commanded you, 'You shall not eat of it,' cursed is the ground because of you; in pain you shall eat of it all the days of your life; thorns and thistles it shall bring forth for you; and you shall eat the plants of the field. By the sweat of your face you shall eat bread, till you return to the ground, for out of it you were taken; for you are dust, and to dust you shall return.'* ▶

Thinking Biblically, where there are thorn fossils found in any sedimentary layer anywhere in the world, these thorns as well as the sedimentary layer had to come into existence after Adam and Eve brought sin into the world, which occurred after the seven-day creation week. Thus, all fossilized thorns and the sedimentary layers in which they are found can only be at most thousands of years old.

Biblical Reason 12 Summary: The Biblical teaching on thorns and the perspective that fossilized thorns support evolution over billions of years are completely incompatible.

⌘ Biblical Reason 13 ⌘
View of Death

If the days of creation are interpreted to be long ages (millions or billions of years) and life supposedly began millions or billions of years ago in a primordial soup, then evolution supports a position that death and suffering had to be in the world before Adam's sin. But Genesis chapter three teaches us that Adam's fall and the arrival of sin and death into the world occurred *after* the seven days of creation. Having death exist anywhere before or during the seven creation days contradicts multiple points and verses in the Bible. Ultimately, death occurring anywhere before or during the creation week undermines the truth that sin is what brought death into a once-perfect creation and that all people need a Savior from sin and death. Additionally, placing death into the world before Adam's sin strikes at the significance that the Lord Jesus Christ conquered death by His crucifixion and resurrection.

It has been suggested that the death of animals does not count as Biblical death, and therefore, the idea that death existed before Adam and Eve's fall into sin can be harmonized with the Bible. But this idea does not actually harmonize with the clear teaching of the Bible. For example, in the Old Testament, there are verses that teach us that the death of animals is sufficient as a sacrifice for sin, so the Bible does indeed identify animal death as real death.

At the end of the sixth day, God looked at everything He had created and pronounced it "very good". To suggest that the six days in Genesis chapter one represent long ages and that the earth existed for billions of years before the six days implies there had to be death and suffering during these six long ages, which directly contradicts God stating at the end of the sixth day that everything He created was "very good". Thus, the suggestion of billions of years becomes a direct attack on God's character and His truthfulness in what He says. A God of Truth would not declare that everything He made was "very good" if He had just created everything using death-dependent evolutionary processes.

Believing that death has always existed, and that it is an evolutionary method responsible for the existence of everything, places death into a very good and perfect creation before Adam and Eve ever sinned. Therefore, accepting death before the entrance of sin into the world implies that death is not the penalty for sin and that death is not the last enemy, both of which are taught in First Corinthians chapter 15.

Per the definition of evolution, death is a key ingredient, as many alleged species' advancements occur because of the elimination (death) of one lesser creature for the advancement into another creature. We can contrast this with creation and the Bible. The Bible teaches that Jesus Christ is the Creator and Sustainer of all life and that death is not a "life advancer" but an *enemy* that will someday be destroyed!

First Corinthians 15:26 states:

◀ *The last enemy to be destroyed is death.* ▶

The Bible is very clear that from man came death (Romans 5:12 and First Corinthians 15:21), but evolution holds that from death came man. Romans 5:12 states:

◀ *Therefore, just as sin came into the world through one man, and death through sin...* ▶

First Corinthians 15:21 states:

◀ *For since by man came death...* ▶

Thus it is impossible for both the six days of life-giving historical events described in Genesis and evolution's dependence upon death over billions of years to be true.

Biblical Reason 13 Summary: The Biblical view that death resulted from Adam's sin which occurred after the seven-day creation week and evolution's view that death has been occurring over billions of years are completely incompatible.

⌘ Biblical Reason 14 ⌘ Future Perfection

Nobody can deny that our earth is far from perfect today, as we all see everywhere the effects of sin, death, suffering and disease. Scripture clearly teaches that at the end of our current earth's history, a new heaven and earth will exist that will again be perfect (See Revelation 21). This new perfect heaven and earth will be similar to the original perfect creation described in Genesis chapter one because it too will not have any sin, death, suffering, or disease. Genesis 1:31 describes the original perfect creation, which includes the original earth, at the end of sixth day:

◀ *And God saw everything that He had made, and behold, it was very good. And there was evening and there was morning, the sixth day.* ▶

Revelation 21:1-5 describes the future new perfect heaven and earth that will be without sin, death, suffering and disease:

◀ *Then I saw a new heaven and a new earth, for the first heaven and the first earth had passed away, and the sea was no more. And I saw the holy city, new Jerusalem, coming down out of heaven from God, prepared as a bride adorned for her husband. And I heard a loud voice from the throne saying, 'Behold, the dwelling place of God is with man. He will dwell with them, and they will be His people, and God Himself will be with them as their God. He will wipe away every tear from their eyes, and death shall be no more, neither shall there be mourning, nor crying, nor pain anymore, for the former things have passed away.' And He who was seated on the throne said, 'Behold, I am making all things new.' Also He said, 'Write this down, for these words are trustworthy and true.'* ▶

By simple definition, anything that the Lord God Almighty declares as "very good" would not include sin, death, suffering and disease. If evolution was true then no future perfection is possible. Why? According to evolutionary beliefs, the original, current, and the future conditions of the earth all include sin, death, suffering, and disease. These evolutionary beliefs directly contradict the plain and clear teaching of the Bible.

Biblical Reason 14 Summary: Evolution's view of the earth's original condition and its future condition completely contradicts the Biblical view of the earth's original condition and its future condition.

⌘ Biblical Reason 15 ⌘
Perspective of All New Testament Writers

One form of compromise regarding the book of Genesis occurs when people say it is just an analogy or a framework for expressing a big picture of creation, but it is not actually about real, historical events and people. One of the hallmarks of correctly understanding scripture is the guideline "scripture interprets scripture". In other words, the Bible must be consistent with itself, and we can use the truth of scripture in one book to help us understand its truth in other books.

The writers of the New Testament books help us understand that the book of Genesis details real history, people, and events. Every New Testament writer (who of course wrote under the inspiration of the Holy Spirit) actually refers to at least one of the first 11 chapters of Genesis, and in each case, the writer acknowledges these early chapters of Genesis as real history describing real people experiencing real events. In 18 of the New Testament books, Genesis chapters 1 to 7 and 9 to 11 are referenced at least once.

Here are a number of places where New Testament writers reference the early chapters of Genesis and the real people and historical events recorded in those early chapters:

- The Apostle Paul refers to Adam in Romans 5:12: ◀ *...transgression of Adam.* ▶
- The Apostle Paul refers to Adam in First Corinthians 15:21: ◀ *For as by a man came death...* ▶ and ◀ *For as in Adam all die...* ▶
- The Apostle Paul refers to Eve in Second Corinthians 11:3: ◀ *...as the serpent deceived Eve...* ▶
- The Apostle Paul contrasts Adam with Jesus in First Corinthians 15:45: ◀ *The first man Adam became a living being; the last Adam became a life-giving spirit,* ▶ and noting: ◀ *...the first man was from the earth, a man of dust; the second man is from heaven.* ▶
- The Apostle Peter references the global flood of Noah in First Peter 3:20: ◀ *...when God's patience waited in the days of Noah, while the ark was being prepared, in which a few, that is, eight persons, were brought safely through the water.* ▶
- The Apostle John references Cain and Abel in First John 3:12: ◀ *We should not be like Cain, who was of the evil one and murdered his brother.* ▶

- Jude references the cities of Sodom and Gomorrah in Jude 1:7: ◀ *...just as Sodom and Gomorrah and the surrounding cities...* ▶
- Jude references Moses in Jude 1:9: ◀ *But when the archangel Michael, contending with the devil, was disputing about the body of Moses...* ▶
- Jude references Cain in Jude 1:11: ◀ *For they walked in the way of Cain...* ▶
- James refers to Abraham in James 2:21: ◀ *Was not Abraham our father...* ▶
- The writer of Hebrews references Cain and Abel in Hebrews 11:4: ◀ *By faith Abel offered to God a more acceptable sacrifice than Cain...* ▶
- The writer of Hebrews references Enoch in Hebrews 1:5: ◀ *By faith Enoch was taken up...* ▶
- The writer of Hebrews references Noah in Hebrews 11:7: ◀ *By faith Noah, being warned by God concerning events as yet unseen...* ▶
- The writer of Hebrews references Abraham in Hebrews 11:8: ◀ *By faith Abraham obeyed...* ▶

Biblical Reason 15 Summary: No writer of any New Testament book considers the first 11 chapters of the book of Genesis to be merely an analogy or framework, but rather all the writers, writing under the inspiration of the Holy Spirit, recognize Adam and Eve as real people and the book of Genesis as a recording of real events and real history.

⌘ Biblical Reason 16 ⌘
Old Testament Verses

In addition to the verses in the first Genesis chapters, additional Old Testament verses teach us about the creation event, how the Lord created everything, and the age of the universe.

Exodus 20:9-11 identifies that Genesis chapter one must be referring to actual 24 hour days, as it would make no sense for God to give an example of His work over billions of years as a model for the Israelite's seven-day week. These verses read:

◀ *Six days you shall labor, and do all your work, but the seventh day is a Sabbath to the Lord your God. On it you shall not do any work, you, or your son, or your daughter, your male servant, or your female servant, or your livestock, or the sojourner who is within your gates. For in six days the Lord made heaven and earth, the sea, and all that is in them, and rested on the seventh day. Therefore the Lord blessed the Sabbath day and made it holy.* ▶

It is also significant to note that these verses in Exodus 20:9-11 teach us that God rested, not that God "is still resting". By saying God rested, which is past tense, these verses teach us that the creation days are now complete, not still occurring in some form of million or billion-year-long "day".

Exodus 31:12-17 also identifies that the Israelite seven-day week has its basis in the seven days of the Genesis creation week. Exodus 31:12-17 confirms that Genesis chapter one must be referring to actual 24 hour days, because just like Exodus 20:9-11 above, it would make no sense for God to give an example of His work over billions of years as a model for the Israelite's seven-day week. The verses read:

◀ *And the Lord said to Moses, 'You are to speak to the people of Israel and say, 'Above all you shall keep my Sabbaths, for this is a sign between me and you throughout your generations, that you may know that I, the Lord, sanctify you. You shall keep the Sabbath, because it is holy for you. Everyone who profanes it shall be put to death. Whoever does any work on it, that soul shall be cut off from among his people. Six days shall work be done, but the seventh day is a Sabbath of solemn rest, holy to the Lord. Whoever does any work on the Sabbath day shall be put to death. Therefore the people of Israel shall keep the Sabbath, observing the Sabbath throughout their generations, as a covenant forever. It is a sign forever between me and the people of Israel that in six days the Lord made heaven and earth, and on the seventh day he rested and was refreshed.'* ▶

Jeremiah 10:12 teaches us that the Lord Jesus created everything by His power and wisdom:

◀ *It is He who made the earth by His power, Who established the world by His wisdom, and by His understanding stretched out the heavens.* ▶

Jeremiah 32:17 teaches us that God made everything by His great power, and making everything was easy for Him:

◀ *Ah, Lord God! It is You who have made the heavens and the earth by Your great power and by Your outstretched arm! Nothing is too hard for You.* ▶

Proverbs 3:19 teaches us that the Lord Jesus' wisdom and understanding enabled Him to build the earth and the universe:

◀ *The Lord by wisdom founded the earth; by understanding He established the heavens.* ▶

Psalm 33:6 teaches us that the Lord Jesus Christ spoke everything into existence. It is not consistent to believe that the Lord spoke everything into existence, which is an action with immediate results, and then also believe that He needed billions of years to perform His creation work:

◀ *By the word of the Lord the heavens were made, and by the breath of His mouth all their host.* ▶

Psalm 104:24 teaches us that all the creatures on the earth were created by Him:

◀ *O Lord, how manifold are Your works! In wisdom have You made them all; the earth is full of Your creatures.* ▶

Psalm 19:1-4 encourages us to remember the Lord God Almighty's glory and handiwork when we see created beauty around us, whether we look during the day or night:

◀ *The heavens declare the glory of God, and the sky above proclaims His handiwork. Day to day pours out speech, and night to night reveals knowledge. There is no speech, nor are there words, whose voice is not heard. Their voice goes out through all the earth, and their words to the end of the world.* ▶

Job 12:7-10 encourages us to remember, when we admire animals, birds, plants, and fish, that the Lord created and sustains all life, and it is He alone that supplies every single breath of life to all people:

◀ *But ask the beasts, and they will teach you; the birds of the heavens, and they will tell you; or the bushes of the earth, and they will teach you; and the fish of the sea will declare to you. Who among all these does not know that the hand of the Lord has done this? In His hand is the life of every living thing and the breath of all mankind.* ▶

Job: 38:4 reminds us that it was the Lord who built everything, and no human being was there to witness the first five days of the creation week:

◀ *Where were you when I laid the foundation of the earth? Tell me, if you have understanding.* ▶

Nehemiah 9:6 encourages us to remember that the Lord Jesus Christ not only created everything but now also preserves His creation:

◀ *You are the Lord, You alone. You have made heaven, the heaven of heavens, with all their host, the earth and all that is on it, the seas and all that is in them; and You preserve all of them; and the host of heaven worships You.* ▶

Biblical Reason 16 Summary: Complementing the verses in the book of Genesis, additional Old Testament verses confirm that the creation week is not compatible with evolution over billions of years.

⌘ Biblical Reason 17 ⌘ New Testament Verses

In addition to the verses in the first Genesis chapters, additional New Testament verses teach us about the creation event, how the Lord created everything, and the age of the universe.

In John 3:12, Jesus states that we should believe Him in both matters of earthly and heavenly things. It is certainly reasonable to apply this to the creation of the earth. The Lord Jesus Christ is the author and inspiration of the entire Bible, and in Genesis chapter one, He describes how He created everything in six days and rested on the seventh day. If we will not believe His words about the creation event but claim we fully believe other heavenly or spiritual things He says elsewhere throughout His word, we are being inconsistent and ultimately not fully believing Him. The verse reads:

◀ *If I have told you earthly things and you do not believe, how can you believe if I tell you heavenly things?* ▶

In Luke 16:31, Jesus quotes Abraham when He warns that we need to receive the words of Moses as truth, remembering it was Moses who wrote Genesis under the inspiration of the Holy Spirit:

◀ *He said to him, 'If they do not hear Moses and the Prophets, neither will they be convinced if someone should rise from the dead.'* ▶

In Mark 10:6 the Lord Jesus states that Adam and Eve were created at the beginning of creation. If each of the days of the creation week somehow represent billions of years, then His statement would make no sense because Adam and Eve would not have been made during the beginning week of time. The point that the very first two people were created during the first week of time only makes sense if the creation days are actual days. The verse reads:

◀ *But from the beginning of creation, 'God made them male and female'.* ▶

John 1:1-3 highlights that all things were made through Jesus the Word and without Him was not anything made that was made:

◀ *In the beginning was the Word, and the Word was with God, and the Word was God. He was in the beginning with God. All things were made through Him, and without Him was not anything made that was made.* ▶

In First Timothy 2:13, the Apostle Paul reminds us that Adam was made first, then Eve. The suggestion that people supposedly evolved from apes or ape-like hominid creatures contradicts this verse because such a suggestion implies that man did not precede a woman and appear first on the Earth. Paul says:

◀ *For Adam was formed first, then Eve.* ▶

Colossians 1:16 teaches us that everything was created by the Lord Jesus Christ, both the things we see and the things we cannot see:

◀ *For by Him all things were created, in heaven and on earth, visible and invisible, whether thrones or dominions or rulers or authorities - all things were created through Him and for Him.* ▶

Hebrews 11:3 highlights that the Lord God created everything out of nothing; even the matter He used to build everything was first created by Him:

◀ *By faith we understand that the universe was created by the word of God, so that what is seen was not made out of things that are visible.* ▶

The Apostle Paul in Romans 5:12 informs us that there were no people existing before Adam. Evolution requires evolving ape-like and hominid-stages of people living and dying for millions of years prior to Adam, which Biblically must be false because there was no sin or death before Adam existed (See Biblical Reason 13). The verse reads:

◀ *Therefore, just as sin came into the world through one man, and death through sin, and so death spread to all men because all sinned...* ▶

The Apostle Paul in Romans 1:20 teaches us that the Lord created the world and everything in it, and we can see His eternal power and divine nature by simply observing His creation:

◀ *For His invisible attributes, namely, His eternal power and divine nature, have been clearly perceived, ever since the creation of the world, in the things that have been made.* ▶

In Acts 17:24-26 the Apostle Paul reminds us that the Lord of Heaven made everything and gives life to all people, according to His timeframes, starting with Adam:

◀ *The God who made the world and everything in it, being Lord of heaven and earth, does not live in temples made by man, nor is He served by human hands, as though He needed anything, since He himself gives to all mankind life and breath and everything. And He made from one man every nation of mankind to live on all the face of the earth, having determined allotted periods and the boundaries of their dwelling place.* ▶

Revelation 4:11, the final book of scripture, concludes that everything was created by the Lord's will, not by any form of random chance:

◀ *Worthy are you, our Lord and God, to receive glory and honor and power, for You created all things, and by Your will they existed and were created.* ▶

Biblical Reason 17 Summary: Complementing the verses in Genesis, additional New Testament verses emphasize that the creation week event is completely incompatible with evolution occurring over billions of years.

⌘ Biblical Reason 18 ⌘
Personal Perspective of the Lord Jesus Christ

If the Lord Jesus Christ speaks of His creation and a young earth everywhere throughout His word, we should not take a position that is contrary to His word. It is impossible to square up evolutionary beliefs spanning billions of years with the Bible, recognizing the Bible gives us the very words of Jesus delivered through the Holy Spirit to the men that penned the books of the Bible.

Think about this powerful question: why do we believe any Bible miracle or event that seems to be supernatural or outside the realm of our physical reality? The answer is faith. We believe God's word by faith. We believe it whether or not we can fully explain it by our own human understanding. We believe it whether or not the Bible's clear teaching is consistent with the popular cultural or scientific beliefs of the time in which we live. In summary, we believe that the Lord Jesus raised Lazarus from the dead, that Noah's flood was global, that Jonah survived three days in a great fish, that Jesus healed sick people with a touch, that the Red Sea was parted for the Israelites to pass through, and that Jesus named every star in our universe, all because of faith in what the Lord has told us in His word, not because we can fully understand and explain these supernatural events solely with our human understanding. We should apply the same faith to understanding the early chapters of Genesis as well as all Bible verses that speak about the creation week and the age of the earth.

It is important to remember that the only eye-witnesses to the creation event were the members of the Trinity: God the Father, God the Son, and God the Holy Spirit. Every human being that has an interpretation or opinion about the creation event was, of course, not there to witness it. Moses was told about it and what he should write in Genesis. This may seem an insignificant point, but it is important to remember this simple point that Genesis is an eye-witness account from the Lord Jesus Christ Himself, relayed through the Holy Spirit to Moses. Just as Genesis begins with "In the beginning", John 1:1-3 echoes another "In the beginning" creation truth:

◀ *In the beginning was the Word, and the Word was with God, and the Word was God. He was in the beginning with God. All things were made through Him, and without Him was not anything made that was made.* ▶

By definition of His character, Jesus can only tell the truth and He will not lie. Thus, what Jesus says in the early chapters of Genesis must be true. John 14:6 assures us that Jesus the Truth will not lie:

◀ *Jesus said to him, 'I am the Way, and the Truth, and the Life. No one comes to the Father except through Me'.* ▶

During His life on earth, when Jesus spoke about the people, accounts, and events of the Old Testament, He treated them as actual historical people, actual historical accounts, and actual historical events, not as figurative allegories or myths. Some of the historical people, accounts, and events that He referred to are highlighted in the following verses:

- Adam and Eve as the first married couple, stated in Matthew 19:3-6:

  ◀ *And Pharisees came up to him and tested him by asking, 'Is it lawful to divorce one's wife for any cause?' He answered, 'Have you not read that He who created them from the beginning made them male and female, and said, 'Therefore a man shall leave his father and his mother and hold fast to his wife, and the two shall become one flesh'? So they are no longer two but one flesh. What therefore God has joined together, let not man separate.'* ▶

- Abel who was murdered, stated in Luke 11:51:

  ◀ *...from the blood of Abel to the blood of Zechariah.* ▶

- Noah and the global flood, stated in Matthew 24:38-39:

  ◀ *For as in those days before the flood they were eating and drinking, marrying and giving in marriage, until the day when Noah entered the ark, and they were unaware until the flood came and swept them all away, so will be the coming of the Son of Man.* ▶

- Moses and the serpent in the wilderness, stated in John 3:14-15:

◀ *And as Moses lifted up the serpent in the wilderness, so must the Son of Man be lifted up, that whoever believes in Him may have eternal life.* ▶

- Moses and the manna from heaven, stated in John 6:32-33:

◀ *Jesus then said to them, 'Truly, truly, I say to you, it was not Moses who gave you the bread from heaven, but my Father gives you the true bread from heaven. For the Bread of God is He who comes down from heaven and gives life to the world.'* ▶

- Moses and his authority to write Genesis, stated in John 5:46-47:

◀ *For if you believed Moses, you would believe Me; for he wrote of Me. But if you do not believe his writings, how will you believe My words?* ▶

- Lot and his wife, stated in Luke 17:28-30:

◀ *Likewise, just as it was in the days of Lot - they were eating and drinking, buying and selling, planting and building, but on the day when Lot went out from Sodom, fire and sulfur rained from heaven and destroyed them all - so will it be on the day when the Son of Man is revealed.* ▶

- The cities Sodom and Gomorrah, stated in Matthew 10:15:

◀ *Truly, I say to you, it will be more bearable on the day of judgment for the land of Sodom and Gomorrah than for that town.* ▶

- Elijah's miracles, stated in Luke 4:25-26:

◀ *But in truth, I tell you, there were many widows in Israel in the days of Elijah, when the heavens were shut up three years and six months, and a great famine came over all the land, and Elijah was sent to none of them but only to Zarephath, in the land of Sidon, to a woman who was a widow.* ▶

- Jonah and the men of Nineveh, stated in Luke 11:32:

◀ *The men of Nineveh will rise up at the judgment with this generation and condemn it, for they repented at the preaching of Jonah, and behold, something greater than Jonah is here.* ▶

Biblical Reason 18 Summary: The personal perspective of the Lord and Savior Jesus Christ, in which He treats the people and events in Genesis as real, actual, historical people and events, is not compatible with evolution and its alleged billions of years of earth's history.

⌘ Biblical Reason 19 ⌘
Authority of the Lord Jesus Christ's Word

Evolution attacks the authority of the words of the Lord Jesus Christ Himself, given to us in the Holy Bible. Here are powerful summary points on why evolution over billions of years is not consistent with the clear and straightforward Biblical teaching about the creation week and the age of the earth.

- The real issue behind understanding the age of the earth is in recognizing the authority of the Lord Jesus, who is Himself identified as the Word. Holding a view that the earth is millions or billions of years old is not being consistent with the book of Genesis and all of scripture. We should never give credence to the scriptural interpretations of anyone, no matter how intelligent they are and no matter what their call or profession is (pastor, theologian, scientist, doctor, or philosopher, for example), when those interpretations contradict God's word. Human opinions that Genesis is not real history should remind us of Satan's first recorded words to Eve, recorded in Genesis: ◀ Did God really say? ▶

- Some will say that it is not important to note all the details mentioned in the book of Genesis, but rather we need only consider the big main points, such as God being the Creator of everything. A simple question logically arises: If the details are not important, why did the Holy Spirit inspire Moses to include these details in His word? Obviously, if details are given to us in the book of Genesis, the Holy Spirit meant them to be there, and we should take note of them. It is not up to sinful people to decide what parts of the Bible are important and true, or not.

- To understand that God the Father, the Lord Jesus, and the Holy Spirit created everything, and to understand approximately when this was done, we need to recognize that only the Bible is truly inspired by the Holy Spirit. Every other source of information that we have - whether science books or theology books or philosophy books or history books and whether written by Christians or atheists or scientists or anyone else - was written by a sinful person who was not inspired in the same way as the men who wrote the Bible were inspired by God the Holy Spirit. Even when considering all kinds of scientific reasons that support the creation week and a young earth, such as those listed in the majority of this book, our ultimate source of truth should be the Bible for understanding the creation event and the age of the earth.

- If the only book available to someone was a Bible, and no other book existed (a hypothetical situation for sure, but to make the point here), someone would never even come up with the idea that the earth is anything other than about

6,000 years old. There are no verses anywhere in the Old Testament or the New Testament that state or imply the earth is millions or billions of years old. All beliefs and interpretations about the earth being millions or billions of years old come from the traditions of men that are outside of scripture.

- Interpreting the seven days of the creation event as long ages (millions or billions of years) is bringing outside ideas into the Bible (eisegesis) rather than letting scripture interpret scripture and receiving the clear grammatical and historical understanding that the verses provide (exegesis). In every instance where someone has not accepted the days of Genesis to be ordinary 24 hour days, they have not allowed the words of the Bible to speak to them in context, as the rules and guidelines of language require for clear communication.

- We need to remember that all scripture is true and inspired by the Holy Spirit, as Second Timothy 3:16 accentuates: ◀ All Scripture is breathed out by God... ▶ The Bible's truth does not depend on our full understanding of the scripture, for its truthfulness is not a relative thing. Rather, it is the Lord's word and is true regardless of how much we understand it or can explain it. A good example is the point that the sun was not created until the fourth day. Some may say that they do not understand how there could be a literal day on creation days one to three without the sun existing. Again, we need to remember what the Bible says about each day is true whether or not we can completely understand it. There is a simple Biblical answer to this question: God created light on the first day, and our sun is not needed for a literal day. What *is* needed for a literal day involves a light source and a rotating earth with a 24 hour revolution cycle. Although scripture does not give us specific scientific details about the light created on the first day, nonetheless, on the first day a light source was created and our rotating earth existed, the only two things needed to produce a literal day.

- The danger of embracing evolution and its billions of years is that it is a step towards positioning sinful people to become the ones to pick and choose which Biblical verses and doctrines to consider true and which verses and doctrines to consider false. This is a direct attack on the sovereignty and character of the Lord Jesus, Who by definition is identified as the Truth. Genesis and many other verses throughout scripture (See Biblical Reasons 17 and 18) teach us that He is the Creator of everything. Evolution attacks His integrity and the veracity of His words.

- If Genesis cannot be understood as historical, actual history, then we have to ask ourselves, "Where in scripture does real history begin?" Many will answer, "With Abraham." But the genealogy of Jesus goes all the way back to Adam (See Luke chapter three). So if we do not start history at Adam but instead with another Biblical person mentioned after Adam, when do we start saying one

Biblical person is really historical and another Biblical person is just an allegory or myth? The point here is that we are not being consistent in our understanding of the Bible when we do not recognize history began on creation week's first day and that Adam was created on the sixth day.

- The only reason to suggest the earth and everything created has an age of millions or billions of years is to accommodate evolution. There are no subjects, themes, topics, or theologies of the Bible that require the age of the earth to be millions or billions of years old in order to be true and historical. Remember that evolution requires billions of years to allegedly allow enough time to change an original first cell into billions of transitional forms and ultimately end up with all living things, including people.

Here are two powerful quotes to help us understand the length of creation days. Theologian Martin Luther, who called the church back to "Sola Scriptura" (Latin for "Scripture Alone") in the 1600s, says:

◀ *When Moses writes that God created Heaven and Earth and whatever is in them in six days, then let this period continue to have been six days, and do not venture to devise any comment according to which six days were one day. But, if you cannot understand how this could have been done in six days, then grant the Holy Spirit the honor of being more learned than you are. For you are to deal with Scripture in such a way that you bear in mind God Himself says what is written. But since God is speaking, it is not fitting for you wantonly to turn His Word in the direction you wish to go.*

The days of creation were ordinary days in length. We must understand that these days were actual days (veros dies), contrary to the opinion of the Holy Fathers. Whenever we observe that the opinions of the Fathers disagree with Scripture, we reverently bear with them and acknowledge them to be our elders. Nevertheless, we do not depart from the authority of Scripture for their sake. ▶

Theologian Charles Spurgeon identifies why scientific beliefs and ideas promoting evolution and its billions of years cannot be trusted:

◀ *We are invited, brethren, most earnestly to go away from the old-fashioned belief of our forefathers because of the supposed discoveries of science. What is science? The method by which man tries to conceal his ignorance. It should not be so, but so it is. You are not to be dogmatical in theology, my brethren, it is wicked; but for scientific men it is the correct thing. You are never to assert anything very strongly; but scientists may boldly assert what they cannot prove, and may demand a faith far more credulous than any we possess. Forsooth, you and I are to take our Bibles and shape and mould our belief according to the evershifting teachings of so-called scientific men. What folly is this! Why, the march of science, falsely so called, through the world may be traced by exploded fallacies and abandoned theories. Former explorers once adored are now ridiculed; the continual wreckings of false hypotheses is a matter of universal notoriety. You may tell where the learned have encamped by the debris left behind of suppositions and theories as plentiful as broken bottles.* ▸

Biblical Reason 19 Summary: Both evolution and its billions of years attack the authority and veracity of the words spoken by the Lord Jesus Himself, given to us in the Bible.

Chapter 11 Endnotes

1 – Frank Sherwood Taylor, "The Ideas and Beliefs of the Victorians: Geology Changes the Outlook", *The Listener*, Vol. 39 (8April1948) pages 573-574.

2 – Michael Denton, *Evolution: A Theory in Crisis* (London: Burnett Books, Ltd., 1985) page 66.

3 – James Rachels, *Created From Animals* (New York: Oxford University Press, 1990) page 5.

4 – W. Peter Gadsby, "Jacques Monod and Theistic Evolution", *Creation*, Vol. 3, No. 2 (May 1980) pages 18-19. <https://answersingenesis.org/theistic-evolution/jacques-monod-and-theistic-evolution>

5 – Julian Huxley, *Issues in Evolution*, Editors Sol Tax and Charles Callender (Chicago: University of Chicago, 1960) page 45.

6 – Frank Zindler, "Debate with William Lane Craig", *Atheism vs Christianity Video* (Grand Rapids, MI: Zondervan, 1996).

7 – Jeremy Rifkin, *Algeny* (New York: Viking Press, 1983) page 244.

8 – G. Richard Bozarth, "The Meaning of Evolution", *American Atheist* (20September1979) page 30.

9 – Josef Ton, *New Life* (15April1982).

10 – G. H. Pember, *Earth's Earliest Ages* (New York: H. Revell Company (1900) page 7.

11 – George S. Hawke, *In Six Days – Why 50 Scientists Choose to Believe in Creation*, Editor John F. Ashton (Green Forest, AR: Master Books, 2001) pages 348-349.

12 – G. Richard Bozarth, "The Meaning of Evolution", *American Atheist* (20September1979) page 19.

13 – James Barr, Letter to David C. C. Watson (23April1984).

14 – Douglas F. Kelly, *Creation and Change* (Scotland, UK: Christian Focus Publications, 1997) page 38.

15 – Ken Ham, *The New Answers Book 1*, Editor Ken Ham (Green Forest, AR: Master Books, 2010) pages 88-112.

16 – Thomas Huxley, "The Lights of the Church and the Light of Science", *Science and Hebrew Tradition*, Vol. 4 of Huxley's Collected Essays (1890) pages 235-236.

17 – Elizabeth Mitchell, "Did Humans Walk Like Chimps Up the Evolutionary Tree?", *Answers in Genesis,* Answersingenesis.org, Pub 15October2015. Web 11January2016. <https://answersingenesis.org/human-evolution/did-humans-walk-like-chimps-up-the-evolutionary-tree>

18 – James J. S. Johnson, *Creation Basics and Beyond*, Editor Jason Lisle (Dallas, TX: Institute For Creation Research, 2013) pages 47-53.

19 – David Catchpoole, "A thorny issue", *Creation Magazine*, Vol. 34, Iss. 3 (July 2012) pages 52-55. <http://creation.com/a-thorny-issue>

Chapter 11 Resources

Order of Creation Events

- https://answersingenesis.org/creation
- *Unformed and Unfilled* by Weston W. Fields

Comparing Worldviews

- http://creation.com/two-worldviews-in-conflict
- http://creation.com/6000-years
- http://creation.com/creation-compromises-questions and answers
- *Did God use Evolution* by Dr. Werner Gitt
- *The New Answers Book 2: Chapter 1* by Answers in Genesis
- *Evolution's Achilles Heel: Chapter 8* by Creation Ministries International

Foundation Attack

- http://creation.com/evolution-hurts-church
- http://creation.com/history-questions-and-answers
- http://creation.com/Did-god-create-over-billions-of-years
- http://creation.com/age-of-the-earth
- https://answersingenesis.org/why-does-creation-matter/crumbling-foundations
- https://answersingenesis.org/evidence-for-creation/the-10-best-evidences-from-science-that-confirm-a-young-earth
- *The New Answers Book 1: Chapter 3* by Answers in Genesis

Naturalism's Failures

- *In Six Days – Why 50 Scientists Choose to Believe in Creation: Chapter 46* by Master Books

Theistic Evolution

- https://answersingenesis.org/theistic-evolution
- http://creation.com/10-dangers-of-theistic-evolution
- http://creation.com/theistic-evolution-what-difference-does-it-make
- http://creation.com/biblical-problems-for-theistic-evolution-and-progressive-creation
- *Old Earth Creationism on Trial* by Master Books
- *Letter To A Theistic Evolutionist* by Icon Publishing Group

Gap Theory

- 💻 https://answersingenesis.org/genesis/gap-theory
- 💻 http://creation.com/the-gap-theory-part-a
- 💻 http://creation.com/the-gap-theory-part-b
- 📖 *Unformed and Unfilled* by Weston W. Fields
- 📖 *The Creation Answers Book: Chapter 3* by Creation Ministries International
- 📖 *Creation and Change: Chapter 5* by Douglas F. Kelly
- 📖 *Creation Basics and Beyond: Chapter 5* by Institute For Creation Research

Writing Style of Genesis

- 💻 http://creation.com/should-genesis-be-taken-literally
- 💻 https://answersingenesis.org/hermeneutics/literary-forms-and-biblical-interpretation
- 📖 *Creation Basics and Beyond* by Institute For Creation Research
- 📖 *15 Reasons to Take Genesis as History* by Creation Ministries International

Clear Meaning of Genesis Chapters 1 and 2

- 💻 http://www.icr.org/article/literal-week-creation
- 📖 *The New Answers Book 1: Chapter 8* by Answers in Genesis
- 📖 *Creation and Change: Chapter* 2 by Douglas F. Kelly

View of People

- 📖 *Did God use Evolution* by Dr. Werner Gitt

Genesis Genealogies

- 💻 https://answersingenesis.org/answers/books/new-answers-book-2/gaps-in-the-genesis-genealogies
- 💻 http://creation.com/genealogy-gaps
- 📖 *Creation Basics and Beyond: Chapter 6* by Institute For Creation Research
- 📖 *The New Answers Book 2: Chapters 5 and 16* by Answers in Genesis

Created Kinds

- 💻 http://creation.com/variation-information-and-the-created-kind
- 💻 https://answersingenesis.org/creation-science/baraminology

Thorns

- 💻 http://creation.com/drawing-power
- 💻 http://creation.com/a-thorny-issue

View of Death

- 💻 http://creation.com/why-death-suffering
- 📖 *The New Answers Book 1: Chapter 26* by Answers in Genesis
- 📖 *Creation Basics and Beyond: Chapter 7* by Institute For Creation Research

Future Perfection

- 💻 http://creation.com/new-earth

Perspective of All New Testament Writers

- 💻 http://creation.com/new-testament-creation

Personal Perspective of the Lord Jesus Christ

- 🕮 *The New Answers Book 1: Chapter 20* by Answers in Genesis
- 💻 https://answersingenesis.org/theory-of-evolution/millions-of-years/jesus-devastates-an-old-earth

Authority of the Lord Jesus Christ's Word

- 💻 http://creation.com/the-authority-of-scripture
- 🕮 *In Six Days – Why 50 Scientists Choose to Believe in Creation* by Master Books

⌘ Chapter 12 ⌘

Why Does It Matter

There is no doubt about it that the doctrine of evolution is the greatest curse in our education system... This philosophy seeks to determine the morale, the principles and practice of virtuous conduct, and to reduce man to the level of animal nature. The surging unrest, the broken homes, the frustrated lives, the increasing divorce rates, the multiplied number of criminals are but the inevitable outcome of the acceptance and practice of this evolutionary doctrine. [3]

Many people feel that believing in evolution over billions of years is not a big deal and that it really does not matter whether anyone believes it or not. Respectfully, what a person believes and takes to heart is a very big deal, because what a person believes shapes everything from how you view yourself and other people to how you view our world's past and future. Here are four powerfully impactful reasons why embracing evolution and billions of years is incredibly dangerous and destructive.

⌘ Impact Reason 1 ⌘
Death, Misery, and Deadly Ethics

Philosopher, cognitive scientist, and atheist Daniel Dennett identifies the brutal reality of the harmful and often deadly impact to people over the last century when military, political, and social leaders not only believe but act out their belief that evolution has occurred over billions of years:

◀ *Evolution is a universal acid, which eats through just about every traditional concept and leaves in its wake a revolutionized worldview. We only have to look at who and what Darwinian ideas have inspired to see the truth of this statement: Marx, Stalin, Mao, Ceausescu, Kim Il-sung, Pol Pot, Hitler, and the eugenics movement (founded by Francis Galton, Darwin's cousin). These offspring of Darwinism murdered over 100 million people and brought untold misery to many more. And we could mention the likes of Australian Peter Singer (now at Princeton University in America where the Hodges and Warfield once taught), who sometimes passes for a 'bioethicist', with his ideas of murdering children and the elderly whose lives are deemed to be not worthwhile… Most atheists hate these connections being pointed out, but they are real.* ▶ [1]

⌘ Impact Reason 2 ⌘
Promotion of Hateful Attitudes

Evolution over billions of years promotes intensely hateful attitudes that are destructive to people and communities within our society. Charles Darwin himself embraced the dangerous belief that people with greater amounts of melanin (the primary pigment that influences our skin color) in their skin are somehow less evolved humans and should be eliminated from society:

◀ *The more civilized so-called Caucasian races have beat the Turkish hollow in the struggle for existence. Looking to the world at no very distant date, what an endless number of the lower races will have been eliminated by the higher civilized races throughout the world.* ▶ [2]

⌘ Impact Reason 3 ⌘ Destructive to Our Society

The teaching of evolution is extremely destructive to our education system, to the individual students in our schools, and to healthy relationships within our society. In his book *Evolution or Creation*, zoologist and creationist Hannington Enoch distinguishes the destructive impact of promoting a belief in evolution occurring over billions of years:

◀ *No wonder that Brigadier General F. D. Frost states in the Fundamentalist, January, 1950, page 21: 'There is no doubt about it that the doctrine of evolution is the greatest curse in our education system.' Whether we read Ward's Dynamic Sociology, or Russell's Code of Morals, or Briffalt's Immoralism or some other book written by the Behaviorist School, they all seem to endeavor to justify and base their conclusions on the bestial nature of man. This philosophy seeks to determine the morale, the principles and practice of virtuous conduct, and to reduce man to the level of animal nature. The surging unrest, the broken homes, the frustrated lives, the increasing divorce rates, the multiplied number of criminals are but the inevitable outcome of the acceptance and practice of this evolutionary doctrine.* ▶ [3]

⌘ Impact Reason 4 ⌘ Eroding Faith in the Lord Jesus Christ

Creationist Ken Ham and statistician Britt Beemer released a book entitled *Already Gone,* which identifies that 60-80% of young church-goers in their teens, twenties, and thirties are leaving the conservative church in which they were raised. Ken Ham and Britt Beemer document many reasons why millions of young people are leaving their faith, but sadly, this tragic reason is included among their list: their pastors and youth leaders often taught evolution and its billions of years as scientific truth. These leaders also taught that evolution and its billions of years could be harmonized with the Bible. A belief in evolution and billions of years undermines the authority of the book of Genesis and eventually the authority of the entire Bible. Ultimately, such beliefs erode faith in Jesus as both Creator and Savior. [4]

⌘ Decision Point ⌘

The bottom line is that the Lord Jesus tells us that He created everything and people are His crowning achievement. Every person, regardless of their education or science background, has to make a decision between embracing evolution and its billions of years or embracing special creation and only thousands of years of history. Former president Ronald Reagan reminds each of us that we need to open our eyes to the Author and Sustainer of all creation:

◀ *I've always believed that we were, each of us, put here for a reason. That there is a plan, somehow - a divine plan for all of us. I know now that whatever days are left of me belong to Him. I also believe this blessed land was set apart in a very special way. A country created by men and women who came here not in search of gold, but in search of God. They would be free people, living under the law with faith in their Maker and their future. Sometimes, it seems we've strayed from that noble beginning, from our conviction that standards of right and wrong do exist and must be lived up to. God, the source of our knowledge, has been expelled from the classroom. He gives us His greatest blessing: life - and yet many would condone the taking of innocent life. We expect Him to protect us in a crisis, but turn away from Him too often in our day-to-day living. And I wonder if He isn't waiting for us to wake up.* ▶ [5]

Chapter 12 – Endnotes

1 – Daniel Dennett, *Darwin's Dangerous Idea: Evolution and the Meanings of Life* (New York: Simon & Schuster, Inc., 1995).
2 – Francis Darwin, *The Life and Letters of Charles Darwin, Volume 1* (New York: D. Appleton and Company, 1888) pages 285-286.
3 – Hannington Enoch, *Evolution or Creation* (London: Evangelical Press, 1966) pages 146-147.
4 – Ken Ham and Britt Beemer, *Already Gone* (Green Forest, AR: Master Books, 2009). Web 24September2015. <https://answersingenesis.org/answers/books/already-gone>
5 – Ronald Reagan, *Address at Annual National Prayer Breakfast* (4February1982). Web 24September2015. <http://www.reagan.utexas.edu/archives/speeches/1982/20482a.htm>

⌘ Appendix A ⌘

List of Scientific and Biblical Reasons

This book provides 77 scientific reasons and 19 Biblical reasons why evolution over billions of years is scientifically and Biblically impossible. As a convenience, here are all the reasons listed together.

Scientific Reason 1 Summary: The suggestion of a Big Bang is not observable, it cannot be scientifically substantiated, and it merely offers hypothetical and philosophical ideas supporting evolution and its alleged billions of years of time.

Scientific Reason 2 Summary: The planets in our solar system consistently follow precise orbits and patterns that can be described by Kepler's three laws of planetary motion. It is scientifically and logically impossible to conclude that random, purposeless, and unintelligent evolutionary processes are responsible for establishing the precise orbits and patterns displayed in our solar system.

Scientific Reason 3 Summary: The existence of spiral galaxies throughout our universe contradicts the supposed billions of years of history. Due to the orbiting motions of individual galaxies and the stars within them, the spiral arms of galaxies would have already dispersed if the universe was billions of years old.

Scientific Reason 4 Summary: Because of scientific observations of the Poynting-Robertson Effect on the space dust within our solar system as well as around other stars within our universe, we can deduce that our solar system cannot be more than 10,000 years old, which contradicts evolution and its claim of billions of years of history.

Scientific Reason 5 Summary: The observed rate of our shrinking sun means that our sun and our earth cannot be millions or billions of years old, contrary to the beliefs essential to evolution and its long age premise.

Scientific Reason 6 Summary: The Jovian planets in our solar system radiate more energy than they are absorbing from our sun, which is opposite to what is expected if they were billions of years old. This scientific observation of our Jovian planets directly contradicts evolution's premise of billions of years.

Scientific Reason 7 Summary: The existence of short-lived comets in our solar system, whose ages do not harmonize with evolutionary timescales of billions of years and for which, as of today, no conclusive evidence exists that they are being replenished from anywhere in our universe, contradicts evolution and its alleged billions of years of history.

Scientific Reason 8 Summary: Scientific Reason 8 Summary: The number of SNRs that should be observed if our galaxy was actually billions of years old do not exist, but rather the number of observed SNRs in our galaxy points to an age of 10,000 or less years. This lack of observed SNRs scientifically contradicts the evolutionary assumption of billions of years.

Scientific Reason 9 Summary: Anything that exists must have a cause, and according to the rules of logic, it must therefore have come from a source. Evolution cannot scientifically explain the source or cause of anything, including all matter.

Scientific Reason 10 Summary: Atoms are an orderly reality of our physical existence that demonstrate remarkable structure, consistency, stability, precision, and interconnectivity, none of which can be the result of the random-chance processes which evolution offers. The design, structure, and behavior of elements, which are beautifully and orderly arranged in the Periodic Table of Elements, cannot be the result of evolution and its unintelligent, random, and goalless processes.

Scientific Reason 11 Summary: All of the unique, finely calibrated, life-giving attributes of water are impossible to scientifically explain as the result of evolution's random and unintelligent processes.

Scientific Reason 12 Summary: It is not scientifically reasonable to suggest that random and purpose-less processes promoted by evolution could have precisely and accurately calibrated all of the universal constants to the specific and exact values required for life to exist on earth.

Scientific Reason 13 Summary: It is not scientifically reasonable to believe that random, designless, chance-based processes promoted by evolution precisely established and continually maintain the consistent laws of physics required for life on earth to exist.

Scientific Reason 14 Summary: It is not scientifically and logically reasonable to believe that evolution's random-chance processes are continually causing an increase in the order, structure, and complexity of living things, because this completely contradicts the Second Law of Thermodynamics.

Scientific Reason 15 Summary: Evolution relies on the idea that our earth is billions of years old, and the formation of stalactites, diamonds, and coal are commonly promoted as supporting geologic processes that require millions of years. Because we now observe that stalactites can be formed in decades and diamonds and coal in less than a year, we now have scientific support that millions of years are not needed.

Scientific Reason 16 Summary: We now have scientific research detailing that granite from all over the world formed in a very short timeframe, which contradicts evolution and its reliance upon billions of years of earth's history.

Scientific Reason 17 Summary: It is not scientifically or logically reasonable to believe that random, designless, chance-based processes associated with evolution over billions of years are responsible for establishing the many precise and specifically calibrated earth attributes contributing to the Anthropic Principle.

Scientific Reason 18 Summary: All available geological data identify that the atmosphere of our earth has always contained significant amounts of oxygen, and thus, has never been able to support an oxygen-free environment required for the alleged spontaneous generation of a first cell.

Scientific Reason 19 Summary: The amount of helium within our atmosphere and the amount of helium within our earth's crust and mantle match the amounts expected for an earth age of about 10,000 years. These scientific observations contradict the expected helium amounts that should be observed if evolution and its long age premise was true.

Scientific Reason 20 Summary: The age of the earth's magnetic field, based on observation and scientific research, cannot be harmonized with evolution and its alleged billions of years of earth's history.

Scientific Reason 21 Summary: The earth exhibits rotational consistency around its axis and orbital consistency around our sun. These attributes of the earth are specifically calibrated at the values needed to support life on earth. It is not scientifically, observationally, or logically reasonable to believe that evolution's random-chance and purposeless processes are the explanation for the earth's finely-tuned, consistent rotation and orbit.

Scientific Reason 22 Summary: It is not scientifically, observationally, or logically reasonable to propose that evolution, based on designless chance-based processes, could be responsible for the complex and interdependent cycles that exist on the earth.

Scientific Reason 23 Summary: The appearance of many geological landscapes on the earth display the results of rapid, aggressive, and catastrophic geologic forces that are not observable today. These landscapes fail to support slow and gradual forces that evolution and its billions of years require.

Scientific Reason 24 Summary: The scientifically observed current rate of mud and sediment accumulation into the Mississippi River Delta cannot be scientifically harmonized with evolution's requirement for billions of years.

Scientific Reason 25 Summary: The rate of sodium accumulation in our oceans provides observable scientific support that is incompatible with the billions of years of earth's history demanded by evolution.

Scientific Reason 26 Summary: The scientifically observed rates of sediment and mud accumulation onto our ocean floors is not compatible with the billions of years of earth history demanded by evolution.

Scientific Reason 27 Summary: Scientific research demonstrates three truths:

1) A mixture of supposedly "advanced" and "primitive" fossilized creatures exist throughout all sedimentary layers all over the earth.
2) There is a complete lack of any conclusive and observable fossilized transitional creatures.
3) There are gaps in the fossil record between all major groups of creatures.

Because all three of these significant scientific observations are true, evolution and its concept that the fossil record displays evolving creatures is scientifically unsubstantiated.

Scientific Reason 28 Summary: If evolution was true, trillions upon trillions upon trillions (See Scientific Reason 64) of transitional human and pre-human fossils would exist. The fact that they do not exist tells us that evolution over long ages is scientifically unsubstantiated and not true.

Scientific Reason 29 Summary: The observable existence of human artifacts found within layers of coal and hardened sandstone reveals that the layers are not really tens or hundreds of millions of years old as proposed by evolution and its dependence upon long ages.

Scientific Reason 30 Summary: The observable existence of the Cambrian explosion within sedimentary layers, in which creatures appear as fully formed fossilized creatures with no accompanying transitional fossils leading up to them and no advanced fossils appearing in the layers above them, displays exactly the opposite of what should be observed according to evolution and its premise of the slow and gradual development of new species through transitional forms.

Scientific Reason 31 Summary: The complete lack of any conclusive scientific evidence of a fossil ancestor for any of the 30 million dinosaur bones discovered to date, which represent about 700 dinosaur species, is not compatible with evolution and its premise that dinosaurs evolved from less advanced creatures living millions of years ago.

Scientific Reason 32 Summary: The observable existence of hundreds of living fossils is not compatible with evolution and its premise that creatures living today evolved from less advanced creatures which lived hundreds of millions of years ago.

Scientific Reason 33 Summary: The existence of polystrate fossils in the fossil record is not scientifically compatible with evolution and its reliance upon time spans of millions of years.

Scientific Reason 34 Summary: We now have scientific evidence that all kinds of objects, whether composed of wood, cloth, or once-living tissue, can petrify in as little as a few years or decades. Therefore, evolution's proposal that petrification supports millions of years is wrong.

Scientific Reason 35 Summary: The observable discoveries of soft tissues within fossils, which at chemical and molecular levels cannot be more than a few thousand years old, scientifically identifies that these fossils are not tens or hundreds of millions of years old as proposed by evolution and its assumption of billions of years.

Scientific Reason 36 Summary: The geologic column fails to scientifically support evolution and its required billions of years of earth history three ways: 1) Observable geological evidence reveals that real-life geologic layers exist in different sequences compared to geologic chart drawings. 2) Geologic layers are missing in 99.6% of the locations where the geologic column is proposed to exist. 3) Fossilized creatures are actually found in other geologic layers contrary to geologic chart drawings.

Scientific Reason 37 Summary: The dating of geologic layers and the fossils within them often results from circular reasoning rather than from real scientific observation and testing. Thus, the geologic dates derived from circular reasoning fail to offer credible scientific support to evolution and its proposed billions of years.

Scientific Reason 38 Summary: The scientific observation that sedimentary rock layers found all over the earth appear very flat and horizontal contradicts what should be observed according to evolution's billions of years. Logically, if these sedimentary layers were really tens or hundreds of millions of years old, due to rain, flowing water, currents, tides, springs welling up, storms, hail, wind, burroughing and digging animals, animals moving across the layers, earthquakes, and plants growing roots, the layers would be full of holes, divots, gorges, gullies, canyons and uneven surfaces.

Scientific Reason 39 Summary: The observable scientific and geologic changes from the eruptions of Mount Saint Helens show that millions of years are not needed to accumulate sedimentary layers, gouge canyons, petrify logs, and form coal layers, all of which evolution promotes took millions of years to form.

Scientific Reason 40 Summary: The scientific observations that oil formation can occur in a few years or decades rather than billions of years and that oil reservoirs have significant pressure which would disperse before billions of years could pass provides scientific support that contradicts evolution's premise of billions of years.

Scientific Reason 41 Summary: Within the Grand Canyon, both the Coconino Sandstone Formation and the Kaibab Upwarp provide geologic examples that had to be formed from rapid and catastrophic water erosion, not from slow and gradual forces. Within the Grand Canyon, there are layers missing from the proposed geologic column. All of these scientific observations are contrary to evolutionary-based expectations that the Grand Canyon was formed over hundreds of millions of years.

Scientific Reason 42 Summary: Because scientific research has shown that radiometric dating cannot accurately date rocks and materials of known age, that radioactive decay rates can be altered with changes in pressure and temperature, and about half of all resulting dates are ignored or never publically announced, radiometric dating cannot be considered a scientifically reliable dating method, and therefore cannot be used to conclusively support evolution and its billions of years of earth's history.

Scientific Reason 43 Summary: Geologists have found that fossils, coal, oil, natural gas, limestone, marble, and graphite from every sedimentary layer of the proposed geologic column contain carbon-14. Because any material that contains carbon-14 cannot scientifically be older than 100,000 years, the sedimentary layers as well as the fossils of once-living plants and animals found within them cannot be millions or billions of years old as promoted by evolution.

Scientific Reason 44 Summary: Based on seven scientifically unverifiable assumptions, carbon-14 dating fails to be a conclusively reliable dating method, and thus, cannot be used to support evolution and an earth age of billions of years.

Scientific Reason 45 Summary: The scientifically observed ratio between the Specific Production Rate of carbon-14 and the Specific Decay Rate of carbon-14 is not in equilibrium within the earth's atmosphere, which is the opposite of what should be observed if the earth was billions of years old.

Scientific Reason 46 Summary: We now know that the Spontaneous Generation of life from non-life is scientifically impossible, which presents a completely unsolvable contradiction to evolution.

Scientific Reason 47 Summary: By definition, evolution requires that life originally arose from non-life. Because this directly contradicts the Law of Biogenesis, a scientific law that has never been falsified, evolution fails to be scientifically true.

Scientific Reason 48 Summary: It is scientifically, statistically, and logically impossible that evolution and its random, purpose-less processes could be responsible for the origin and existence of proteins.

Scientific Reason 49 Summary: It is statistically and scientifically impossible for evolution to have formed, by random chance, the estimated 25,000 enzymes that function in a human body. Evolution cannot circumvent the scientific and statistical impossibility of originally forming an enzyme by random chance.

Scientific Reason 50 Summary: It is biologically, chemically, and statistically impossible for the cells in our body, which include DNA, RNA, proteins, cellular organelles and membranes, to be the result of random, spontaneous chance-based processes. Therefore it is scientifically impossible for a supposed first cell to have formed according to a series of small, successive genetic and physical changes over billions of years as promoted by evolution.

Scientific Reason 51 Summary: The incredible structure, design, and complexity of a bacterium's flagellum screams of design and purpose that cannot be the result of the random chance-based processes proposed by billions of years of evolution.

Scientific Reason 52 Summary: It is scientifically impossible for the fascinating, two-week process of butterfly metamorphosis as well as the butterfly's amazing abilities to be the result of evolution over millions of years.

Scientific Reason 53 Summary: The incredible design, beauty, and capabilities displayed by the 12,000 mile journey of the Artic Tern, the acute sensing ability of the Bombyx Mori beetle, the advanced sonar system in dolphins, the highly sensitive vibration detection system in grasshoppers, the advanced optics in lobster eyes, the heat radiation sensing ability of vipers, and the complex design of Trilobites eyes are all examples of capabilities that the random chance-based processes proposed by evolution over billions of years could never create.

Scientific Reason 54 Summary: The pervasive pattern of discontinuity observed in the appearance and structure of all living things displays exactly the opposite of what should be observed if evolution over billions of years was true.

Scientific Reason 55 Summary: It is impossible for the irreducible complexity demonstrated in living things to be the result of billions of small, continual, successive genetic and physical steps offered by evolution over billions of years.

Scientific Reason 56 Summary: Because the process of photosynthesis involves complicated chemical systems and hundreds of chemical reactions that must occur in a specific sequence, it is an irreducibly complex chemical system which could never be the result of random and accidental causes proposed by evolution over billions of years of earth's history.

Scientific Reason 57 Summary: In seeds, the design, the ability to follow genetic instructions, the ability to use stored energy, and the ability to grow into a fully developed plant with trillions and trillions of specifically purposed cells all demonstrate irreducible complexity, none of which can be scientifically explained by evolution.

Scientific Reason 58 Summary: While the reality of the Fermi Paradox certainly does not by itself scientifically remove support for evolution, it does provide one more example of scientific observations which are consistent with the impossibility of evolution over billions of years.

Scientific Reason 59 Summary: Because there are information attributes which could never arise accidently by random chances, and because greater information must first exist in order to produce subsequent information, and because information requires a three-part condition of sender/meaning/receiver, evolution and its premise that information resulted from purely naturalistic and materialistic causes is logically and scientifically impossible.

Scientific Reason 60 Summary: The existence of the laws of math, which are orderly, intelligent, consistent, universal, absolute, and true, is not compatible with evolution and its premise that random chances and purely naturalistic and materialistic processes are somehow the sole explanation for the source of everything, including all laws of math.

Scientific Reason 61 Summary: The Fibonacci Sequence appears as Golden Ratios in animals, plants, people, and space. The Fibonacci Sequence displays distinctive order, stunning patterns, and incredible beauty, none of which can be consistently explained by evolution's belief that unintelligent, random, and purely naturalistic processes created order, patterns, and beauty in everything.

Scientific Reason 62 Summary: Because of the amount of information stored within one molecule of DNA, and because one double-helix of human DNA is made with six billion specifically sequenced nucleotide bases, and because the building of the cellular machinery needed to read DNA is itself coded by the DNA, and because DNA equidistance relationships display the opposite to what evolution requires, and because the myth of junk DNA has been debunked, and because of the complexity and precision displayed during DNA replication, it is utterly and scientifically impossible for evolution over billions of years to have created the magnificent structure, design, beauty, language, and complexity of DNA.

Scientific Reason 63 Summary: Because DNA mutations are neutral or harmful 99% of the time, and because all mutations never create brand new DNA, and because no scientific experiment has ever observed a mutation creating a brand new gene (or hormone, enzyme, or organ), and because genetic research tells us the human genome could not endure the current amount of observed generational SNVs over millions of years, and because mutations do not mathematically occur frequently enough to support the timeframe required by evolution, DNA mutations scientifically fail to support evolution over billions of years.

Scientific Reason 64 Summary: Calculating the human population size for a hypothetical one million years, by using a significantly more conservative human population growth rate compared to what we observe today, provides mathematical and scientific support that contradicts evolution's claim that people lived and evolved over the last 2.5 million years.

Scientific Reason 65 Summary: The incredible examples of order, design, beauty, and interrelated functionality demonstrated within the human body are impossible to explain by the mere random-chance, directionless, purposeless processes offered by evolution. It is impossible to explain the existence of irreducibly complex body systems involving our speech capability, our brain and nervous system, our ears, our eyes, our heart and circulatory system, our metabolism, our immune system, our reproductive system, and our blood clotting ability as results of small, successive, continual, genetic and physical changes over millions of years, as required if evolution was true.

Scientific Reason 66 Summary: The incredible design, complexity, order and beauty demonstrated by the ATP Synthase enzyme to produce massive amounts of ATP molecules provides an irreducibly complex system that could never be the result of small, successive, intermediate genetic and physical steps offered by evolution over billions of years.

Scientific Reason 67 Summary: The proposal that people have vestigial organs implies that the human genome is losing operational DNA rather than gaining brand new DNA, which is actually the opposite of what evolution requires to be true. Recent scientific research has discovered at least one specific purpose within our bodies for every alleged vestigial organ, giving us another observable and scientific reason why evolution over billions of years is not true.

Scientific Reason 68 Summary: No process, scientific or otherwise, exists for anything non-material to come into existence purely from a material cause, and thus, evolution cannot scientifically explain the origin and existence of everything that is non-material.

Scientific Reason 69 Summary: The idea that the age of our universe is billions of years old flows from a naturalistic assumption which is based upon an unscientific personal bias and is not scientifically testable or supportable.

Scientific Reason 70 Summary: It is statistically impossible for evolution over billions of years to have occurred.

Scientific Reason 71 Summary: Natural selection, by its own definition, cannot be the originator of brand new DNA within a creature and thus, it cannot cause a single cell to evolve into a fully developed plant, animal, or human. Additionally, by its own definition, natural selection would eliminate any intermediate, nonworking structures before they became fully functional. Therefore natural selection scientifically fails to support evolution and its requirement for the development of complex life forms from "simpler" forms.

Scientific Reason 72 Summary: Speciation never creates a new species that has brand new genes, but rather it always results in a reduction in number or the reshuffling of existing genes within a new species. Evolution requires the creation of billions of brand new genes to "evolve" a new creature from an original single cell, but speciation scientifically and genetically fails to offer this capability to evolution.

Scientific Reason 73 Summary: Haeckel's biogenetic law has been genetically, scientifically, and observationally shown to be a fraud, and therefore cannot be used to support belief in evolution.

Scientific Reason 74 Summary: The extinction of dinosaurs is a popular idea used to support evolution over millions of years. But scientifically, because of the recent discovery of many types of still-soft tissues within dinosaur fossils and the known chemical and molecular limitations for tissue decay resistance, we now have scientific research confirming that dinosaurs lived thousands of years ago and not millions of years ago.

Scientific Reason 75 Summary: The idea that homologous structures in living creatures support evolution cannot be conclusively or scientifically substantiated, and recent scientific advancements in embryology and genetic research have confirmed that assumed homologous structures fail to show the required genetic relationships for evolution to be true.

Scientific Reason 76 Summary: Scientific research has not provided the required observations to enable the drawing of a scientifically accurate evolutionary morphological tree diagram, which directly contradicts evolution over long ages.

Scientific Reason 77 Summary: No conclusive scientific evidence exists to support evolution over billions of years.

Biblical Reason 1 Summary: The order of events promoted by evolution are completely contrary to the historical order of events detailed in Genesis chapters one to three.

Biblical Reason 2 Summary: The worldview of the Bible and the worldview promoted by evolution are completely incompatible.

Biblical Reason 3 Summary: Evolution attacks the very foundation of the Bible, which is provided in the book of Genesis starting at chapter one.

Biblical Reason 4 Summary: The belief system of naturalism is a foundational philosophy for evolution that is completely incompatible with the Bible.

Biblical Reason 5 Summary: The proposal of theistic evolution and the attributes of God the Father, God the Son, and God the Holy Spirit are completely incompatible.

Biblical Reason 6 Summary: A clear, straightforward reading of Genesis chapters one to two teaches us that the gap theory and the clear context and meaning communicated in the first two chapters of Genesis are not compatible.

Biblical Reason 7 Summary: The writing style of Genesis is a historical narrative that is not written as poetry or as a parable or as an allegory. Therefore, the seven-day creation week described in the first two chapters of Genesis is not compatible with evolution and its alleged billions of years of earth's history.

Biblical Reason 8 Summary: The straightforward meaning and clear contextual usage of the Hebrew word Yom used in Genesis chapters one and two teach us that the seven days of the Biblical creation week are completely incompatible with evolution and its billions of years of earth history.

Biblical Reason 9 Summary: The Biblical view of the unique and special creation of people is completely incompatible with evolution's view of people.

Biblical Reason 10 Summary: The genealogies and timeframes listed in Genesis chapters 1, 5, 7, and 11 help identify that the earth is about 6,000 years old, and therefore, it follows that evolution and its required billions of years of earth's history are not true.

Biblical Reason 11 Summary: The idea of one common ancestor from which all life descended, as promoted by evolution over billions of years, is completely incompatible with the Biblical teaching of many distinctly created kinds.

Biblical Reason 12 Summary: The Biblical teaching on thorns and the perspective that fossilized thorns support evolution over billions of years are completely incompatible.

Biblical Reason 13 Summary: The Biblical view that death resulted from Adam's sin which occurred after the seven-day creation week and evolution's view that death has been occurring over billions of years are completely incompatible.

Biblical Reason 14 Summary: Evolution's view of the earth's original condition and its future condition completely contradicts the Biblical view of the earth's original condition and its future condition.

Biblical Reason 15 Summary: No writer of any New Testament book considers the first 11 chapters of the book of Genesis to be merely an analogy or framework, but rather all the writers, writing under the inspiration of the Holy Spirit, recognize Adam and Eve as real people and the book of Genesis as a recording of real events and real history.

Biblical Reason 16 Summary: Complementing the verses in the book of Genesis, additional Old Testament verses confirm that the creation week is not compatible with evolution over billions of years.

Biblical Reason 17 Summary: Complementing the verses in Genesis, additional New Testament verses emphasize that the creation week event is completely incompatible with evolution occurring over billions of years.

Biblical Reason 18 Summary: The personal perspective of the Lord and Savior Jesus Christ, in which He treats the people and events in Genesis as real, actual, historical people and events, is not compatible with evolution and its alleged billions of years of earth's history.

Biblical Reason 19 Summary: Both evolution and its billions of years attack the authority and veracity of the words spoken by the Lord Jesus Himself, given to us in the Bible.

⌘ Appendix B ⌘

☙ Resources For Your Own Exploring ☙

Book Resources:

- 📖 *15 Reasons to Take Genesis as History* by Creation Ministries International
- 📖 *Body By Design* by Dr. Alan L. Gillen
- 📖 *Creation and Change* by Christian Focus Publications
- 📖 *Creation Basics and Beyond* by Institute For Creation Research
- 📖 *Creation Facts of Life* by Dr. Gary Parker
- 📖 *Darwin's Demise* by Dr. Joe White
- 📖 *Did God use Evolution?* by Dr. Werner Gitt
- 📖 *Dragons: Legends and Lore of Dinosaurs* by Master Books
- 📖 *Evolution Exposed* by Answers in Genesis
- 📖 *Evolution Impossible* by Dr. John F. Ashton
- 📖 *Evolution the Grand Experiment* by Dr. Carl Werner
- 📖 *Evolution's Achilles Heel* by Creation Ministries International
- 📖 *Guide to Creation Basics* by Institute For Creation Research
- 📖 *Guide to Dinosaurs* by Institute For Creation Research
- 📖 *In Six Days – Why 50 Scientists Choose to Believe* in Creation by Master Books
- 📖 *In The Beginning* by Dr. Walt Brown
- 📖 *In The Beginning Was Information* by Dr. Werner Gitt
- 📖 *Letter to a Theistic Evolutionist* by Dr. Duane T. Gish
- 📖 *Old-earth Creationism on Trial* by Tim Chaffey and Dr. Jason Lisle
- 📖 *On the Seventh Day* by 40 Authors and Edited by Dr. John F. Ashton
- 📖 *Refuting Evolution* by Dr. Jonathan Sarfati
- 📖 *The Book of Beginnings* by Dr. Henry M. Morris III
- 📖 *The Creation Answers Book* by Creation Ministries International
- 📖 *The Genius of Ancient Man* by Don Landis
- 📖 *The Geology Book* by Dr. John D. Morris
- 📖 *The Global Flood* by Dr. John D. Morris
- 📖 *The Illustrated Origins Answer Book* by Paul S. Taylor
- 📖 *The Lie: Evolution* by Ken Ham
- 📖 *The New Answers Book 1* by Answers in Genesis
- 📖 *The New Answers Book 2* by Answers in Genesis
- 📖 *The Young Earth* by Dr. John D. Morris
- 📖 *Thousands...Not Billions* by Dr. Don DeYoung
- 📖 *Unformed and Unfilled* by Weston W. Fields
- 📖 *Untold Secrets of Planet Earth - Dire Dragons* by Vance Nelson
- 📖 *Untold Secrets of Planet Earth - Flood Fossils* by Vance Nelson

Online Resources Emphasized in this Book:

- Answers in Genesis
  - https://answersingenesis.org
- Biblical Geology
  - http://Biblicalgeology.net
- Center for Scientific Creation
  - http://www.creationscience.com/onlinebook
- Creation Ministries International
  - http://creation.com
- Institute For Creation Research
  - http://www.icr.org
- Scientific Facts & Evolution, including the Evolution Cruncher
  - http://evolutionfacts.com

Online Supplemental Resources:

- All About Creation
  - http://www.allaboutcreation.org
- Creationism.Org
  - http://www.creationism.org
- Creation Facts
  - http://www.creation-facts.org
- Creation Moments
  - http://www.creationmoments.com
- Creation Research Society
  - https://www.creationresearch.org
- Facts of Creation
  - http://www.truesciencefacts.com
- Faith Facts
  - http://www.faithfacts.org
- Genesis Apologetics
  - http://genesisapologetics.com
- The True Origin Archive
  - http://www.trueorigin.org

Index

Adam (from Bible) 168, 171, 172, 175, 180, 182, 183, 184, 185, 187, 188, 191, 192, 194, 197, 198
Ager, Derek 63, 71
Akridge, Russell 23
Amino acid 102, 103, 104, 105, 114, 160
Anthropic Principle 46, 47
Archaeomagnetic data 49
Archaeopteryx 159
Arctic Tern 109
Asimov, Isaac 38, 139
Atmosphere 34, 46, 47, 48, 49, 51, 55, 56, 82, 93, 95, 104
Atom 16, 32, 33, 34, 35, 36, 41, 49, 51, 96, 102, 105, 110, 148
ATP (Adenosine Triphosphate) 105, 114, 142, 143
ATP Synthase 142, 143
Bacteria 102, 107, 114, 129, 133, 141
Barr, James 176
Beck, William 154
Beemer, Britt 206
Begley, Sharon 162
Behe, Michael 106, 113
Bergman, Jerry 69, 156
Biblical worldview 169, 170, 171, 172
Big Bang 16, 17, 18, 20
Blood clotting 112, 141, 142
Bock, Walter 154
Bombyx Mori beetle 109, 111
Boornazian, Aram 23
Bozarth, G. Richard 172, 174
British Coal Measures 71
Brown, Walt 22
Butterfly 70, 108, 109
Cambrian explosion 67, 68, 69
Capra, Fritjof 134
Carbon-14 dating 91, 93, 94, 95, 96
Carson, Ben 142
Cathedral of Florence, Italy 128
Cell 47, 48, 72, 73, 102, 104, 105, 106, 107, 114, 115, 128, 129, 130, 134, 139, 140, 141, 142, 143, 150, 151, 152, 153, 160, 161, 198
Chambered Nautilus 127
Charlesworth, Brian 134
Chlorophyll 114
Chloroplasts 114
Coal 39, 67, 71, 82, 83, 84, 93, 94
Comets 20, 21, 24, 25
Coconino Sandstone Formation 85, 86
Cohen, I. L. 151
Colorado River 52, 53
Columbia River Basalt Group 53
Commonwealth Scientific and Industrial Research Organization, Australia 84
Conklin, Edwin 150
Cook, Melvin 48
Crab Nebula 25, 26
Created kinds 168, 183
Creation events 168
Creation week 172, 175, 176, 177, 178, 179, 182, 184, 185, 189, 190, 191, 193, 196, 198
Crick, Francis 128
Cycle 34, 46, 51, 52, 197
Darling, David 31
Darwin, Charles 9, 11, 13, 63, 65, 68, 100, 105, 113, 140, 161, 162, 171, 178, 205
Davies, Paul Charles William 36
Da Vinci, Leonardo 128
Dawkins, Richard 68, 159
Death 9, 62, 100, 141, 168, 169, 170, 172, 174, 175, 179, 184, 185, 187, 189, 192
Dembski, William 150
Dennett, Daniel 205
Denton, Michael 129, 156, 161, 170
De Beer, Gavin 156
DeSalle, Rob 90
DeYoung, Don 102
Diamond 39, 93
Dimroth, Erich 48
Dinosaur 69, 70, 73, 153, 155, 159, 168
Dolphin 110, 111
Dragon 155
Druyan, Ann 25
Dudley, H. C. 92
Dunham, David W. 23

Dyson, Freeman 101
Eddy, John A. 22, 23
Ehrlich, Paul 12
Einstein, Albert 123
Einstein's Gulf 123
Eisegesis 172, 197
Element 32, 33, 34, 47, 91, 92
Embryology research 156, 157
Encyclopedia Britannica 80
Engel, A. E. J. 161
Enoch, Hannington 206
Enzyme 103, 104, 130, 132, 134, 141, 142, 143
Evolution definition 9
Evolution events 168
Evolution unknowns 11
Evolution worldview 169
Exegesis 172
Fermi, Enrico 116
Fibonacci, Leonardo Pisano 126
Fibonacci Sequence 126, 127, 128
First Law of Thermodynamics 16, 17
Ford, Dwaine L. 102
Fossil 53, 62, 63, 64, 65, 66, 67, 68, 69, 70, 71, 72, 73, 79, 80, 81, 93, 94, 95, 155, 157, 158, 159, 183, 184
Future perfection 186
Galaxy 20, 25, 26, 46, 127
Gap theory 175
Genealogies in Genesis 176, 180, 181, 182
Gentry, Robert 40
Geologic column 79, 86, 93, 94
Gish, Duane 12, 64
Gitt, Werner 124
Goldberger, Robert F. 105, 161
Golden Angle 126, 127
Golden Rectangle 126
Golden Spiral 126, 127
Gould, Stephen Jay 54, 63, 66, 68, 152
Grand Canyon 52, 53, 83, 85, 86
Granite 40, 41
Grasse, Pierre Paul 133
Grasshopper 110, 111
Gravity 17, 19, 32, 35, 37, 47
Great Pyramid of Giza 128
Green, David Ezra 105, 161
Gulf of Mexico 55
Guth, Alan 37
Haeckel, Ernst 153, 154
Haeckel's Biogenetic Law 153, 154
Ham, Ken 206
Harold, Franklin 104
Harwit, Martin 20
Haskins, Caryl 130
Hawke, George S. 173, 174
Helium 48, 49
Heribert-Nilsson, Nils 64
Heylmun, Edgar 54
Histidine amino acid 104
Hodge, Andrew 141
Homologous structures 156, 157
Honey Bee 127
Hoyle, Fred 36, 100, 101, 150
Human brain 139, 140, 142
Human ear 140, 142
Human eye 140, 142
Human heart 140, 142
Human immune system 140, 141, 142
Human reproductive system 141, 142
Human speech 139, 142
Hummingbird 110
Humphreys, David Russell 26, 49
Huxley, Julian Sorell 171
Huxley, Thomas 151, 158
Information 38, 66, 106, 115, 123, 124, 125, 128, 129, 130, 131, 132, 151, 153, 154, 156, 196
Intermediate form 63, 64, 112, 115, 152
Irreducible complexity 34, 106, 112, 113, 114, 115, 139, 141
Issus Coleoptratus insect 109
Javor, George T. 52
Johnson, Philip 12
Jones, J. S. 66
Jupiter 24
Kaibab Upwarp 85, 86
Kelly, Douglas 133
Kenyon, Dean H. 160
Kepler, Johannes 19
Kimberley, Michael 48
Kitts, David B. 63
Kuiper Belt 25
Law of Biogenesis 101, 134
Laws of math 125
Laws of physics 17, 19, 25, 35, 37
Leakey, Mary 158
Lee, Robert E. 91
Lewis, C. S. 148

Lewontin, Richard 12
Libby, Willard 96
Light year 25
Lincoln Memorial, Washington D.C. 39
Living fossil 70, 71
Lobster 70, 110, 111
Loewenstein, Jerald 157
Loowit Canyon, Oregon 83
Luther, Martin 198
Lisle, Jason 106
Magnetic field of earth 46, 49, 50, 92, 95, 109
Mammal skeleton 70
Milky Way Galaxy 25, 26, 46, 127
Mississippi River delta 55
Milton, Richard 108
Mitochondria 143
Moon 19, 46, 51
Monod, Jacques 171
Morphological tree 157, 158
Morton, Jean Sloat 103
Moses (from Bible) 176, 182, 188, 189, 191, 194, 195, 196, 198
Mount Ngauruhoe, New Zealand 91
Mount Saint Helens, Oregon 82, 83, 91
Museum of Stone, Siberia 39
Mycoplasma bacterium 102
Natural Selection 68, 102, 132, 140, 151, 152, 153, 161, 171
Naturalism 12, 149, 173, 174
Neptune 24
Newton, Isaac 19, 36, 37, 149
Nutation 51
Ocean sediment 56, 57
Ogden, J., III 92
Oil 83, 84, 85, 93, 94
Oller, John W., Jr. 123
Olsson, I. U. 92
Oort Cloud 25
O'Rourke, J. E. 80
Owen, Richard 155
Parthenon, Greece 128
Pasteur, Louis 100
Patterson, Collin 66, 158, 159
Patterson, Roger 85
Pember, G. H. 173
Periodic Table of Elements 33, 34
Photon 21
Photosynthesis 46, 113, 114
Phyla 67, 111, 112, 133
Pineapple 127
Pinecone 126
Pitman, Michael 132
Planets 19, 20, 24, 46, 116, 149, 179
Plants 31, 35, 47, 51, 54, 62, 81, 82, 83, 91, 94, 114, 126, 127, 128, 156, 168, 179, 183, 190
Popper, Karl 129
Poynting, John Henry 21
Poynting-Robertson Effect 21, 22
Probability 102, 103, 150, 151
Protein 47, 72, 102, 103, 104, 105, 107, 114, 129, 130, 141, 143, 160
Purdom, Georgia 153
Pure Grown Diamonds, New York 39
Rachels, James 171
Radioisotopes and the Age of The Earth (RATE) project 94
Radiometric dating 83, 91, 92, 93, 94
Raup, David M. 63
Reader, John 65
Reagan, Ronald 207
Red blood cells 72, 73, 106, 153, 155
Reducing atmoshpere 47, 48
Riddle, Mike 103
Ridley, Mark 63
Rifkin, Jeremy 38, 154, 172
Robertson, Howard Percy 21
Rose, Michael R. 12, 157
Sagan, Carl 25
Saint Peter Sandstone Megasequence 53
Samec, Ron 24
Sarfati, Jonathan 25, 96
Save-Soderbergh, Torgny 92
Saturn 24
Schweitzer, Mary 73, 155
Scientific method 11, 13, 80, 149, 159
Second Law of Thermodynamics 16, 17, 38, 39
Sedimentary rock 48, 53, 62, 63, 64, 67, 68, 69, 70, 71, 80, 81, 82, 83, 85, 86, 93, 94, 183, 184
Seed 113, 115, 126, 127
Sequoyah Caverns, Alabama 39
Simpson, George Gaylord 11, 62, 154
Single Nucleotide Variants 132
Singularity 16, 31
Slusher, Harold 22

Smith, Wolfgang 13, 161
Snelling, Andrew 40, 83, 84
Sodium 55, 56
Soft tissues 72, 73, 155
Solar system 19, 21, 22, 24, 25, 32, 46, 148, 178
Speciation 153
Spetner, Lee 132
Spirit Lake, Oregon 82
Spontaneous generation 48, 100, 101
Spurgeon, Charles 198
Supernova 25, 26
Sunderland, Luther 66
Sunflower 126, 127
Stalactites 39
Stanley Miller experiment 160
Statistical impossibility 102, 103, 104, 107, 138, 141, 150, 151
Step Canyon, Oregon 83
Sun 19, 21, 22, 23, 24, 46, 50, 51, 95, 104, 109, 149, 168, 197
Supernatural 149, 171, 173, 174, 193
Tahmisian, N.J. 161
Taylor, Frank Sherwood 170
Theistic evolution 174, 175
Thompson, Keith S. 154
Thorn 183, 184
Thornton, Joseph W. 102
Threshold of possibility 103, 150
Todd, Scott C. 149
Tomkins, Jeffrey 106, 130, 160
Ton, Josef 173
Trilobite 111
Tyrannosaurus Rex 73, 155
Uniformitarianism 52, 54
Universal constants 35, 36
Uranus 24
U.S. Supreme Court 160
Vardiman, Larry 48
Vestigial organs 143, 144
Viper 110, 111
Volition 124
Wagner, Gunter 156
Wald, George 13
Walker, Tas 91
Water 32, 34, 35, 41, 46, 47, 51, 52, 53, 55, 56, 72, 81, 82, 83, 85, 86, 104, 114, 168, 175, 181, 187
Watson, James 128
Watson, Lyall 65
Werner, Carl 70, 71
West, Ronald 80
Whipple, Fred 21
White, Tim 66
Whitehead, Gordon 140
Wigner, Eugene Paul 125
Wise, Kurt 111
Worldview 169, 170, 171, 172, 205
Woodmorappe, John 40, 79
Yom (Hebrew) 177, 178, 179
Young, Edward J. 176
Yosemite National Park, California 53
Zihlman, Adrienne 157
Zindler, Frank 171
Zircons 49
Zumberge, James H. 54

Made in United States
North Haven, CT
05 November 2026

21772142R00124